MW01180862

TAB Books
Division of McGraw-Hill, Inc.
New York San Francisco Washington, D.C. Auckland Bogotá
Caracas Lisbon London Madrid Mexico City Milan
Montreal New Delhi San Juan Singapore
Sydney Tokyo Toronto

Hopped-Up Harleys and Performance Street Machines

Carl Caiati

The words Harley, Harley-Davidson, H-D are registered trademarks. This publication is not affiliated with Harley-Davidson. Words, model names, designations, and so on mentioned herein are official Harley-Davidson trademarks and are used for identification purposes only. This is not an official Harley-Davidson publication.

FIRST EDITION
FIRST PRINTING

TAB Books is a division of McGraw-Hill, Inc.

Library of Congress Cataloging-in-Publication Data
Caiati, Carl.
Hopped-up Harleys and performance street machines / by Carl Caiati.
p. cm.
Includes index.
ISBN 0-8306-4394-X
1. Harley-Davidson motorcycle—Customizing. I. Title.
TL448.H3C347 1993 93-43604
629.227'2—dc20 CIP

Acquisitions editor: April D. Nolan
Editorial team: Joanne M. Slike, Executive Editor
Gradon Mechtel, Editor
Stacey R. Spurlock, Indexer
Production team: Katherine G. Brown, Director
Jana L Fisher, Layout
Linda L. King, Proofreading
Donna M. Gladhill, Quality Control
Design team: Jaclyn J. Boone, Designer
Brian Allison, Associate Designer
Cover design: Sandra Blair Design, Harrisburg, Pa.
Cover photo: Arlene Ness "Sled."

HT1
4386

Acknowledgments

A book of this magnitude, or mini-magnitude if you prefer because you could write volumes on Harley hop-ups, could not have been undertaken without the assistance of friends, cohorts, and certain respected personalities of the Harley-Davidson performance field.

One cannot write a true performance book without including Carl Morrow, of Carl's Speed Shop, the "Guru of Go." Carl and his son Doug hold more Harley land-speed and racing records than most, and maybe all, of their contemporaries, and they keep adding more records to their roster every year. Carl can get more out of a Harley engine than anyone I have ever known. His motors are famous for their precision craftsmanship, power engineering, and ability to hold together. Carl was gracious enough to provide more than his fair share of time in technical areas and also the expertise that makes this book worthwhile reading for the casual or serious performance-minded street rider.

Though we have both been in the custom field for the same three decades or so, I had never met Arlen Ness formally till about two years ago. We shared the acquaintance of the late Larry Kumferman, former editor of *Cycle Guide* and *Custom Chopper*, who always regarded Arlen as the very best and never got tired of telling me so. While Arlen was going about building his own world and enhancing Harleys with his one-off special creations, I was spawning a reputation in motorcycle writing and custom painting.

I had the good fortune to meet Arlen at Daytona in 1993 at Crane Cam's open house in the company of his son Cory, a real chip off the old block. Both entertained me with bike lore and Arlen allowed me to photograph his 1987 FXR, un-dresser, convertible, bike-within-a-bike, a classic tour de force. The bike is featured on the cover. It is real cover material, as many bike mags were quick to realize. A creative genius, Arlen is most amiable, not to mention a true gentleman in every sense, and a stellar attraction at any bike function.

Rivera's Mel Magnet, another gracious gentleman, gave unstinting assistance and much sound advice as to what quality performance is all about. He also provided many necessary parts and photos. He not only assisted me with Rivera's line, but also with the products of his competitors that he felt were worthy. That is what is meant by unstinting assistance.

John McGee, of Bandit Machine, was also most cooperative in providing me with material on his first-rate clutch systems. Tom Rudd of Küryakyn contributed greatly by allowing me to feature his parts. His Hypercharger is a great air/filter induction accessory. Tom Motzco of Drag Specialties provided me with some necessary photos of his product line. Ken Smith of S&S, as always, provided assistance and gratefully so, since S&S produces so many crucial power train and engine components needed for this book.

"Monster Motor Man" John Sachs, a Florida neighbor and excellent Harley engine builder, was there crafting, modifying, assembling, and *flow-bench* testing while I shot away with my camera.

I adore Betty Dostie of Crane Cams. She is a great representative for her company, and as technical coordinator, was always available when I needed her. Betty and Crane gave me great backup. My only complaint about Betty is that she's married. But that had a redeeming factor as well. Her old man, Jerry, has a Harley and he did extensive testing on the new Accel Thunder-Twin fuel-injection system, which was essential to this book.

The mention of Accel reminds me of the valuable assistance I received from them at Daytona and their great rep, Ben Kudon, who first introduced me to the Accel fuel-injection system. Ben assisted greatly in providing tech data on the injection system and many other Accel items, as well as providing actual parts.

Ken Rogers, of Dyna Ignition, was another valuable contributor to this book, providing necessary data and parts from their stock of the finest ignition systems and coils produced for Harley engine performance applications.

Many thanks also to Bill Chappell of Sputhe Engineering—Sputhe was there all the way—as well as Steve Storz of Storz Performance Accessories and Kirk Van Scoten of Sumax Inc. Kirk has always been a great provider, ready and willing to assist to the best of his availability. Jay-Brake and their staff were a great asset. Alan Mahan of Delkron I could not have done without.

The shops in my local area also deserve some thanks. Felix Lugo's Phantom Motorcycles of Ft. Lauderdale, Florida, provided parts and photography. Jake Jacobowski's Custom Accessories was willing to share some of head mechanic Scott Barringer's expertise and assembly know-how.

My ace buddy Sandy Roca was there providing me with data from his files, sound advice, and a host of line drawings for this book.

John Andrews gets a nod for cam photos and material (parts, etc.), all which were welcome and necessary for this book.

Ron Dickey of Axtell went all the way providing all that I needed as well as some crucial performance and the theoretical data so necessary to a book of this type.

Ron Paugh from Paughco has been working with me for ages. He goes back as long as I do in this business and has always provided necessary parts and assistance on virtually every big project in the past 30 years. Ron is one of the real good guys, and his custom and stock parts are always the finest.

Last but not least my thanks to my buddy Pete "the Greek" Dotoratos. As well as a good friend, Pete is a whiz with Harley engines. Critical and exacting, he spent many a late hour with me in his garage assembling, shooting photos, and involved in the grind connected with putting together a technical book. He was unfailing in his availability and assistance. His wife Adele also deserves my thanks and appreciation for doing some fine under-the-gun typing of this manuscript.

CONTENTS

FOREWORD

If they gave out academic accolades in modern motorcycle mechanics, Carl Caiati would surely hold a doctorate degree in Harley-Davidson. He has spent a great portion of his half-century life researching, wrenching, and writing about Americ's last—and greatest—motorcylce marque. If there is a way to modify, transmogrify, or otherwise make this line of venerable V-Twins perform beyond their already-muscular capabilities, Carl has done it. He has worked with some of the most knowledgeable Harley tuners and builders in the world—including Carl Morrow and Scott Baringer. He's discovered and catalogued a universe of performance information from some of the best minds in motorcycle-dom.

Ever since Carl began combining his lightning-fast grasp of engine technical theory with elucidating photography and easy-to-understand writing, the entire two-wheeled reading public has been the better for it. Carl's technical features have been published worldwide in the pages of magazines like *Popular Mechanics, Supercycle, American Iron, and Hot Bike*. He's explored a full range of subjects—from specific how-to's to step-by-step project bikes to comprehensive technical overviews. And, he's painstakingly perfected not one but two of his own Harleys from clean-sheet-of-paper ideas to a brace of fast-moving and functional steel steeds—a Sportster and a Big Twin.

In no small part thanks to Carl, Harley motorcycle mechanics has moved from the dirty-fingernail, back-room, bare-knuckle bravodo of the early years to the sophisticated tuning science that it is today. Over the years, he's helped to draw back the veil of black magic of engine and chassis modification to make it understandable and achievable for all.

Now at long last, all of Carl's technical talents and breadth of knowledge are showcased in one convenient place—the pages you are about to read. So kick back, enjoy, and profit from the technical treasures he's about to impart. I think we can all stand the strain.

Sandy "Jake" Roca
Contributing Editor
Motorcycle Collector, Superbike, and Hotbike

Harley riders and enthusiasts of today give almost as much consideration to upgrading engine performance as to customizing and beautifying the appearance of the overall bike itself. Optimum performance is a desirable goal among Harley owners and riders. Though the basic Harley Evo engines are excellent performers out of the box, their performance potential can be dramatically increased with simple to major hop-up procedures, many of them within the scope of the neophyte or weekend mechanic. This is due to proliferation of quality aftermarket goodies made available specifically for early- and late-model Harley engines by today's leading accessory manufacturers.

With today's improvements we see a marked difference in the persona of the contemporary rider as opposed to the rider of yesterday. We have electric starting replacing the older kick starter, roads are more refined and traversable, and the suspension systems of Harleys are vastly improved. Because of better riding conditions riders are demanding more overall power, a better-performing, hotter street machine within reasonable parameters.

The questions most Harleyphiles ponder are: what kind of riding will I be doing, how much performance gain will I realize, and how dependable and trouble-free will it be? According to S&S Products, a leader in Harley engine performance parts, the contemporary rider fits into one of five categories: the hardcore rider, the enthusiast, the weekend warrior, the hobbyist, and the racer. This book is aimed at the first four rider groups. Racers are a breed apart and get involved with advanced technology and state-of-the-art procedures that are not conducive to engine longevity.

The hardcore and the enthusiast riders utilize their bikes for daily transportation, averaging about 10,000 miles of yearly road travel. Their main concern is reliability, but in many instances they toy with the idea of better performance and more horsepower. They are prone to be romanced into better or improved carburetors, a power-bolstering exhaust system, and maybe a mild improved-performance cam. Weekend warriors limit riding time to shorter runs: they are part-time riders but also competitive. Durability is not a major problem; though a concern. The weekend warrior wants to keep up or have a slight edge over his or her rivals. The hobbyist is also a part-time rider of sorts, not overly concerned about the mechanics or theory of engine makeup, but he or she, too, wants more go.

Why do our aforementioned Harley riders consider and implement performance parts? Hardcore and enthusiast riders might require a bit more power for safety reasons, like passing or handling heavier load conditions. The weekend warrior might want that competitive edge. Many riders upgrade their powerplants for their own ego trips or just the fun of experiencing an uplifting change. When rebuilding their engines, many owners will add upgrade goodies, since their engines have to be rebuilt anyway. It can cost the same or even less to rebuild an engine with performance rather than stock parts. If the cost is the same and the parts are equal or of better quality, the case for upgrading becomes stronger. Readers of this book are advised to ask themselves what their performance requirements are. The changes incorporated will determine how much horsepower and torque gains are realized and how dependable and trouble-free the gains will be.

All traditional Harley engines—Big-Twins, Sportsters, Evolutions, and Evolution Sportster twins—leave much room for improvement in all areas. All the Harley mills respond to *stroking*, or increasing the stroke. They also respond beneficially to higher lift cams, bigger pistons, *porting*, and *polishing*, as well as more involved refined headwork, and better carburetion setups.

Almost all the parts and kits from current manufacturers are designed to be used with or to augment stock components with minimal modification to the original engine parts themselves. Some major hop-up procedures do, however, require modification by knowledgable mechanics or machinists. Many products are designed so that the average motorcycle mechanic can install them and some installations or changes are simple enough even for the neophyte.

Many modifications can be made to the Harley engine, any Harley engine, and substantial performance gains realized without sacrificing dependability.

The bulk of the material in this book is devoted to the newer Evolution engines, the Big-Twins predominantly, since the major part of the performance market is geared toward these engines. However, most of the principles and procedures apply to Sportsters and older or vintage V-Twins. I trust that the information will be of value to all Harley owners who wish to eke out a little more go from their street machines.

CHAPTER 1

Chassis & frame modifications

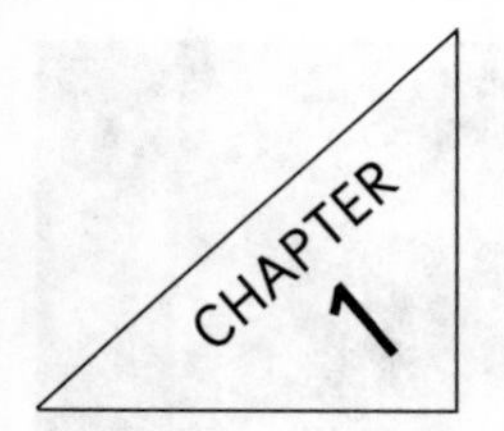

WHEN DEFINING performance, neophytes seldom give proper thought to chassis and frame geometry. One key factor in motorcycle performance and handling is the frame and related components, which can affect the speed and cornering characteristics of the street or competition Harley.

The Harley frame, any Harley frame, is at best a compromise, and a poor one at that, when one considers the improvements built in by the major accessory frame manufacturers of today. The greatest detriment with stock Harley frames is weight. The offerings of such manufacturers as Paughco, Arlen Ness, Sumax, California Pro Street Chassis Works, and others, provide us with a host of alternatives. These frames are lighter (some up to ⅓ lighter than stock), more rigid, offer the option of mounting better suspension components, incorporate reconfigured steering geometry, and are of better structural quality. Today's street rider has available handling performance once found only in the realm of serious roadracers.

Chassis geometry

Optimum handling and performance require neutral steering. It must be quick and responsive, but not oversensitive, uniform but not heavy. Frame and chassis designers integrate rake and trail angles, offset, modified wheelbases and weight distribution. These are all variables that when properly coordinated combine to create neutral steering characteristics. The *rake angle* is that at which the steering head is positioned relative to the angle of true vertical. The more the steering head or neck tube is tilted back, the more increased the rake angle, and therefore, less handlebar movement is required to affect movement at the wheel. Steep rakes also minimize fork seal sliding friction (in telescopic forks), which also allows the use of shorter wheelbases and helps to transfer more of the motorcycle's weight to the front end.

The *caster*, or steering angle, is the degree of angle that the steering head has been displaced from the vertical. On most motorcycles, rake and caster angles are the same, since fork tubes are parallel to the steering head. Since the steering head is connected to the fork tubes via the triple clamps, an increase in steering head angle will also increase trail. *Trail* is the distance between the point where the line of pivot on the steering head axis meets the pavement and a vertical line drawn through the front axle and center of the front tire's contact patch. The more trail, the more pronounced the self-centering effect, and the more inherent stability. Trail assists in re-centering a turned front wheel. With little trail, bikes tend to be less stable at high speeds. Too much trail, on the other hand, imparts heavy steering loads that inhibit quick turning and detract from stable cornering.

Offset is the distance the axle is placed in front of the steering axis and also helps to determine the trail distance. Short offsets increase trail while larger offsets decrease trail.

The offset also determines where the mass of the front end is in relation to the steering axis. A more centralized steering mass will respond more effectively to handlebar manipulation (Fig. 1-1).

Wheelbase is also a factor contributing to handling: it affects rake, trail, and the center of gravity. One must remember that raking the neck or extending the front end also increases the wheelbase. Wheelbase may also be lengthened by extending the rear of a frame or, on a sprung frame, by implementing a longer swing arm. Though a long wheelbase will increase stability, it will create a greater turning radius and promote slower handling. In essence, the properties of a longer wheelbase will reinforce the characteristics of added rake and trail, but stretching a wheelbase beyond 72 inches will make a Harley extremely difficult to maneuver and jockey around (Tables 1-1 and 1-2).

In the halcyon days of motorcycle frame modification, guys had to handbuild their own chassis by trial and error. They had to rake, stretch, and modify their frames themselves or take them to a local frame specialist. Today that's a thing of the past made obsolete by the proliferation of custom frames available in the motorcycle aftermarket. With modern technology and miracle metals the manufacturers offer top-quality products. The leading frame manufacturer Paughco Inc. offers over a dozen Harley frame styles. Any rake or stretch configuration you can come up with is available, with over 500 variations. All the Paughco frames are fabricated of drawn-over-mandrel, .120-inch wall steel (Figs. 1-2, 1-3, and 1-4).

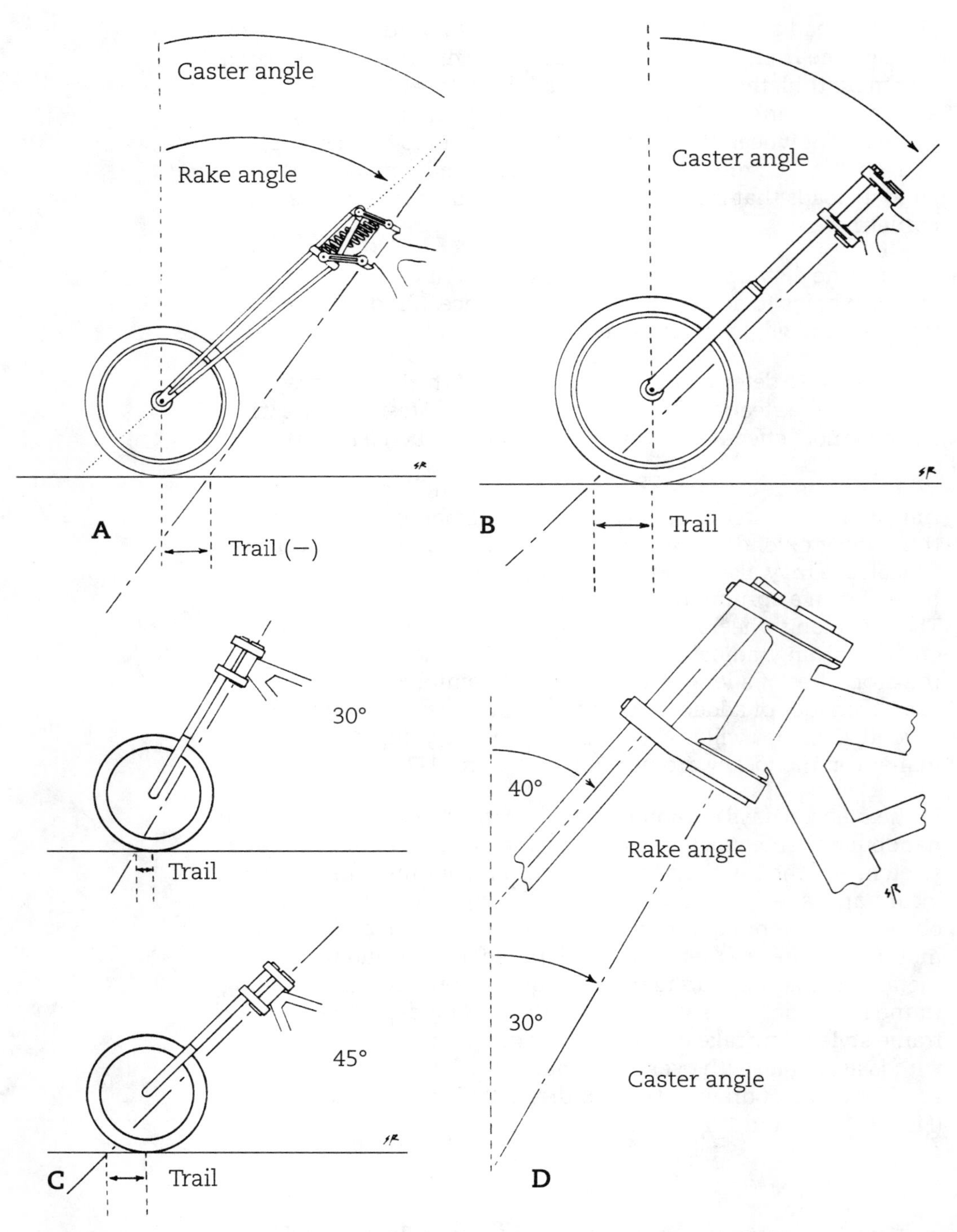

1-1 *The effects of rake, trail, and caster angle. Increasing the steering head angle increases the trail. Caster angle also affects trail geometry. When the fork is tilted back, the axis of steering slants ahead of the motorcycle.*

Table 1-1 The effects of chopper modifications on frame geometry

	Incr. caster angle	*Incr. rake angle*	*Incr. trail*	*Lower center of gravity ground clearance*	*Same C/G & G.C.*	*Higher C/G & G.C.*	*Reduced trail*	*Longer wheelbase*
Cut & bent steering neck	X	X	X	X				X
Raked neck & extended forks	X	X	X		X	or X		X
Extended forks only	X	X	X			X		X
Rake plate		X		X			X	X
Raked triple trees		X		X			X	X
Altered girder links		X		X			X	X
The above trio with extended forks	X	X	X		X	or X		X
Longer rocker plates (on springer forks)							X	X
Bigger front wheel	X	X	X			X		
Smaller rear wheel	X	X	X	X				
Longer than stock rear swing arm or hardtail								X
Shorter rear shocks	X	X	X	X				

Table 1-2 The effects of altered frame geometry on handling characteristics

	Heavy low speed steering *High-speed stability* *Slow chassis reaction* *More cornering effort* *Under-steer*	*Neutral*	*Easier maneuverability* *High-speed lightness* *More responsiveness* *Less cornering effort* *Over-steer*
Increased rake angle	X		
Increased caster angle	X		
Increased trail	X		
Lower center of gravity			X
Higher center of gravity	X		
Reduced rake and trail			X
Longer wheelbase	X		

1-2
The Paughco S151-5 five-speed Evolution frame will replace stock frames and is available in 5-degree, 10-degree; or custom rakes.

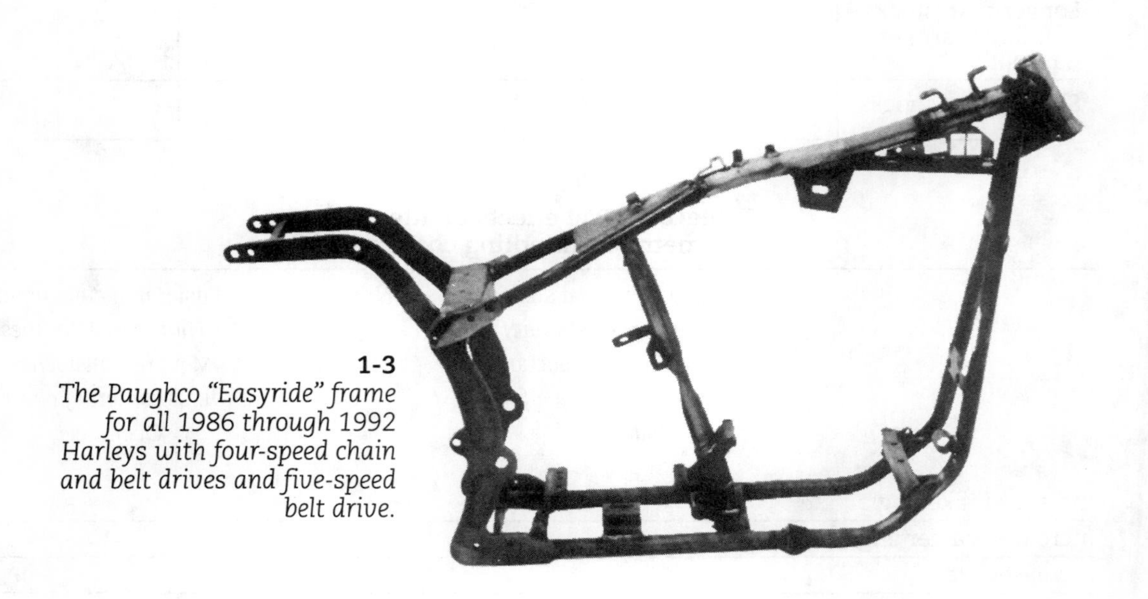

1-3
The Paughco "Easyride" frame for all 1986 through 1992 Harleys with four-speed chain and belt drives and five-speed belt drive.

1-4
The Paughco rigid-wishbone in 5-degree and 10-degree rakes. Four- and five-speed chain or belt drives are available with the five-speed W-218A "tranny" plate.

Arlen Ness, who has been building serious motorcycle parts since before Harley had electric starters, offers a truly custom performance frame featuring ⅞-inch diameter, .188-inch wall, 4130 *chrome moly* tubing, a tried and proven material, both light and strong (Fig. 1-5).

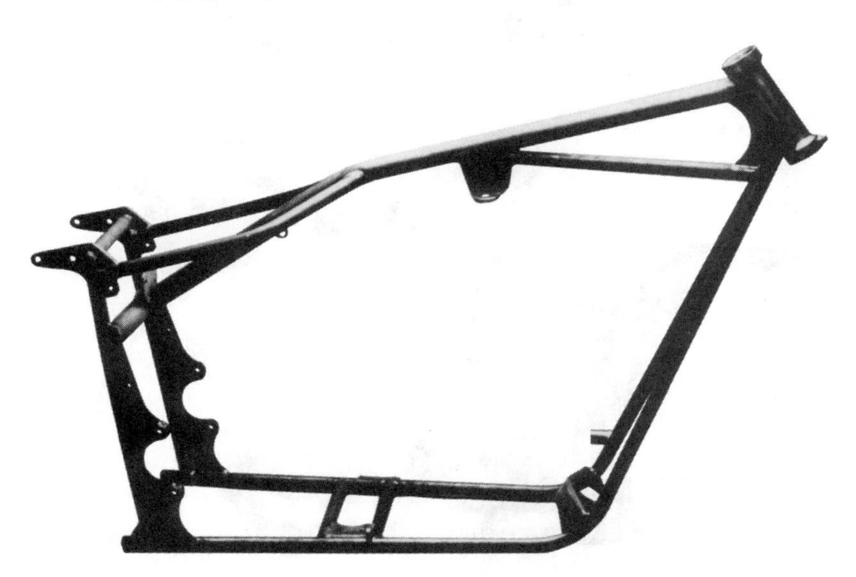

1-5
The Arlen Ness custom 5-speed Evo frame is light and durable and comes complete with wrap-around oil tank.

Sumax, another leader in quality aftermarket parts, has an array of frames in their catalog. They offer replacement frames for any year, "softail," including varied rake and stretch and rubber engine mounts. They even offer a unique softail-rubbermount version for Sportsters, welcomed by fans who desire the softail look (Figs. 1-6 and 1-7).

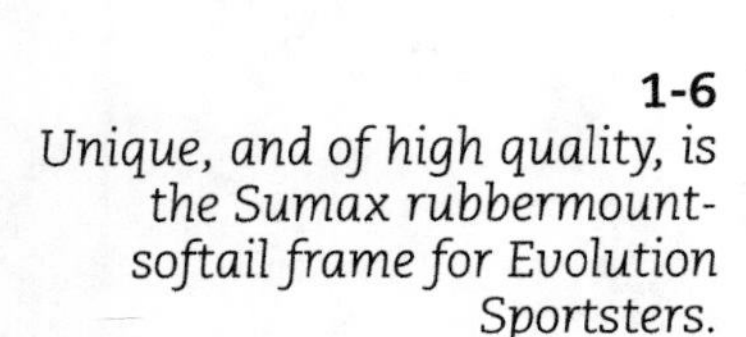

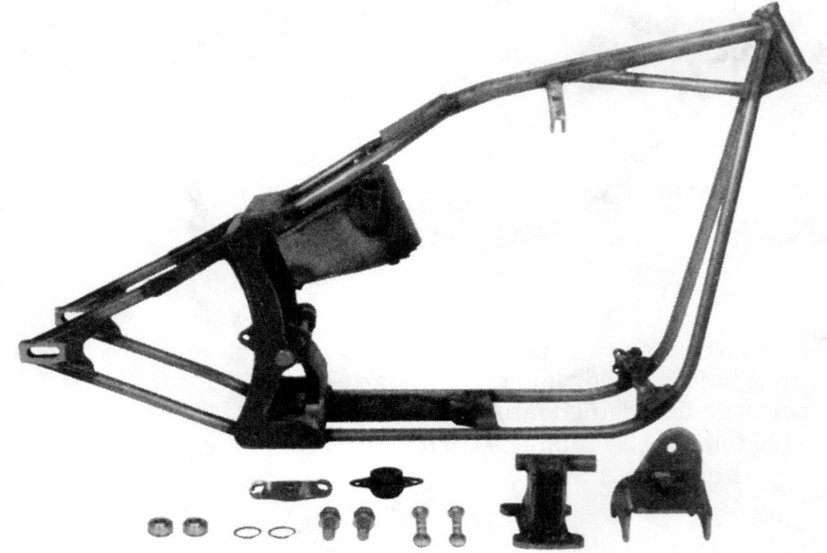

1-6
Unique, and of high quality, is the Sumax rubbermount-softail frame for Evolution Sportsters.

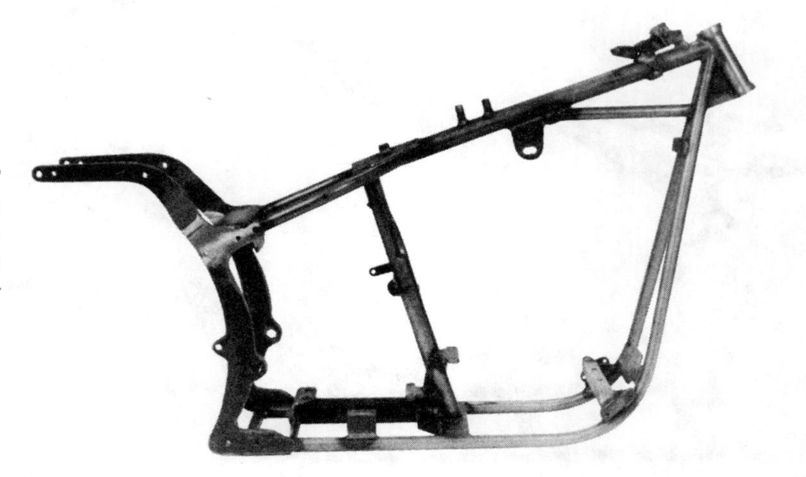

1-7
The popular Sumax FXST universal replacement frame for Evo or Shovelhead motors with four- or five-speed transmissions. Designed for 1986 and later softail frames.

Custom Chrome Inc. offers a wide selection of frames that will allow the performance street-bike builder to build anything from early replicas, to choppers, to FXR rubber-mounted lookalikes. They even market a Big-Twin lookalike early Sportster swing arm frame for 1957 through 1981 models.

California Pro Street Chassis Works offers all-out FXR custom chassis that are precision-made. Their FXR custom chassis uses .095-inch wall, DOM, tubing with 1.250-inch cradle and 2-inch backbone tube sizes. Neck rake angles are offered in 30.5-degrees and 33.5-degrees or can be special-ordered per customer request. The frames are *MIG* welded in state-of-the-art fixtures and weigh in at 45 pounds. Their frames are mild to wild according to customer specifications.

Chopper Guys features Big-Twin rigid frames to accommodate Evo four- or five-speed models with either belt or chain drive. For precision alignment, they use fully machined solid billet steel pieces, in place of cast steel parts. These frames are *heli-arc* welded of .125-inch wall, 1020, DOM tubing. Unfortunately, no Sportster frames are available yet.

Front ends

When choosing a front end, it should complement the frame and the performance characteristics you wish to incorporate. A front end must perform well but should look good, too. You wouldn't want a Wide-Glide on a chopper nor would you want a narrow Ceriani on a Street-Glide. However, save for some Harley stockers, most front ends couple good looks with functionality.

The rake of the frame is a prime consideration when choosing a front end. If your frame's neck rake is 3 or 4-degrees, you will need a relatively short front end. If your frame is radically raked a short front end will impart a nose-down look. A longer front end would be mandatory. When opting for the long look, you might want to consider a *springer*. They are offered in stock to 18 inches over, if you should so require. On long front ends, a springer can provide better damping action.

Springers incorporate an external damping system working on springs, similar to the conventional shock absorber. A springer is a bracket-mounted device, which transfers road action up the length of the struts to a pair of (or single) springs mounted in the crown of the unit. In operation, over street irregularities, the springs compress against a loading stud, which retains the upper spring loop in a retaining collar. The springs compress against the collar offering resistance and hence damping. Because the springer transfers road input up the length of its struts, a portion of the road shock is absorbed as the struts lift into the collars. Because springers are less effective than telescopic forks, a greater stress load is transferred to the crown, which transfers it to frame and, in turn, to the rider as vibration. So much for springers (Figs. 1-8, 1-9).

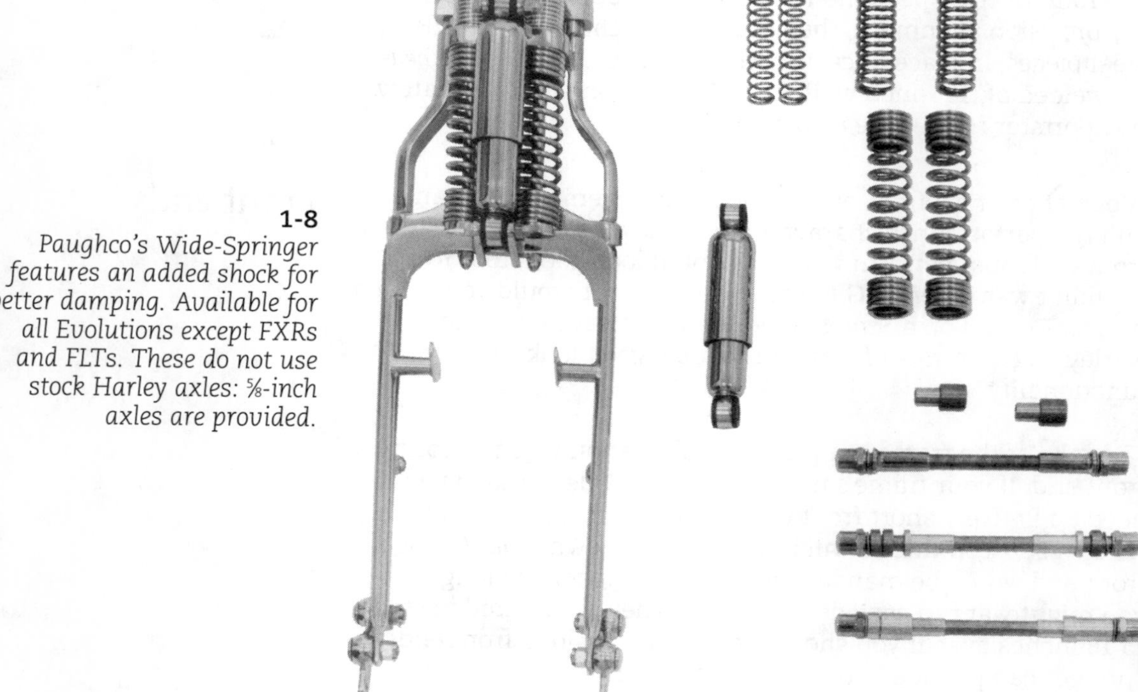

1-8
Paughco's Wide-Springer features an added shock for better damping. Available for all Evolutions except FXRs and FLTs. These do not use stock Harley axles: ⅝-inch axles are provided.

1-9
The "rockers," on all conventional springers, transfer the road input up through the struts to the crown springs, which provide the damping.

Telescoping forks are a more efficient and comfortable way to go and their telescopic spring action does much to dampen road input vibration. They also offer optimum handling characteristics, particularly in cornering. In telescoping fork applications, the fork tubes transfer front-wheel motion through the crown, but the internal long springs and damping oil create a better damping action. Mildly extended fork tubes, marketed by "Forking by Frank," yield a slight rearward center of gravity contributing to a change in weight transfer angle as well. With increased fork lengths you might consequently have to change the rear suspension as well, since the added weight transfer will cause increased pre-loading at the rear when the rear section is shock loaded.

There are numerous aftermarket telescopic front forks that are unequivocally state-of-the-art. White Brothers markets quality forks in both racing and road versions and features an effective upside-down version proven to be most effective on FXR Harleys.

Forcelle-Italia, formerly known as Ceriani, offers what are probably the best telescopic forks, having enjoyed great popularity for over two decades.

These forks are marketed and distributed by Storz Performance, which offers more than one version of these Italian state-of-the-art units. Chrome-moly tubes, machined aluminum sliders, and billet aluminum triple clamps make these forks not only attractive but also lightweight. Fixed damping, quick-change fork seals, and hard-anodized damper rods are standard on the Forcelle-Italia fork. One very popular version is the Forcelle-Italia Wide-Glide, designed for Harleys that will accept the stock Wide-Glide front hub or the narrow Sportster or FX hub, if spacers are used. Storz is also the exclusive distributor for replacement Forcelle-Italia parts for Harley 35- or 39-millimeter forks. Storz replacement forks have 42-millimeter-diameter tubes and are a direct bolt-on replacement, 7 pounds lighter than the stock Harley units.

The newest addition to the Storz line is the Storz-Forcelle-Italia upside-down front fork (Fig. 1-10). Extremely popular with the Sportster set, these forks offer the utmost in rigidity and damping action. The upside-down fork, actually a male slider or inverted fork (narrow at the bottom and wide at the top), is much stiffer than a conventional fork and doesn't tend to flex as much due to the larger outside diameter. Its outer top tube diameter is 54 millimeters, as opposed to the late-style stock Harley's 39 millimeters.

1-10
The Storz-Forcelle-Italia upside-down fork mounted on a Sportster. This fork provides increased stiffness and improved damping action.

The increased stiffness, along with improved damping action, translates into superior handling. The forks feature fixed damping, with an air valve on top of each leg for increasing the spring rate, if desired. This fork kit is designed for Evo Big-Twins and Sportsters but will fit some earlier models as the 1983 XR-1000 also. The forks will accept either stock Harley or Performance Machine brake calipers, and each lower leg has mounting lugs so that dual-disc brakes can be mounted.

Fork oil

The proper selection of fork oil will aid in maximizing the damping abilities of telescopic forks, stock or aftermarket. Adding more fork oil will reduce the volume of air in the fork tube. As the fork is compressed, the lesser volume of air will promote an increase in internal pressure, which acts in the same manner as stiffer springs. With the springs out of the tubes and the tubes bottomed out, oil is added to fill the tube to within ⅝ inch from the top. The tubes are then extended and the springs reinstalled. Another effective way to increase fork damping is to use a heavier oil with a viscosity rating between 15 and 30.

Shock absorbers

The function of shock absorbers is to isolate rider and machine from the bumps and irregularities of the road, or *road input*. With proper shocks the rider can control the bike more easily. With improper suspension the bike is not totally safe, and uncomfortable to boot. Shocks, like front ends, incorporate a spring and a damping device.

Each shock is designed for a specific bike or to incorporate specific functions. These functions are relative to the spring rate, damping characteristics, and the leverage exerted by the swing arm. Swing arm, or softail Harleys, utilize shock absorbers for rear suspension. The spring rate of a shock absorber determines how much "bounce" a spring can absorb. When the springs are fully compressed, with the coils touching, coil bind occurs. To help overcome this, most contemporary, mechanical shock absorbers are extension damped.

After the spring has absorbed the initial big shock it tends to rebound. Without a damper the spring would continue to bounce uncontrollably. The damper inhibits rebound energy and converts it to heat, which is absorbed by the internal shock, oil and springs. The leverage exerted by the swing arm is determined by the locations of the swing arm pivot point and the shock mounting point. The amount of swing arm travel is designed to achieve a proper shock-ride characteristic. Most shocks adjust for pre-load by changing the amount of spring travel. Heavier motorcycles offer a better ride than smaller ones because the heavier bike has more mass, or weight, and more resistance to inertia. Lighter suspension components and wheels also provide better ride characteristics, since lighter components react more quickly and responsively to road input.

Most aftermarket shocks are vastly superior to Harley OEM components. Stock Harley FXST shocks, for example, cause their owners to register many complaints. One is that the stock FXST tends to *wallow* when pushed hard in fast turns. Rider comfort and cornering stability suffer with the stock units. By far, the best softail replacement shocks are the Fournales, Oleopneumatic units, available from Sumax, Inc. These billet aluminum creations feature unrivaled shock damping action, are virtually impossible to bottom out, and provide a softer ride. This shock increases damping as the unit compresses. As the linear length of the shock changes the resistance doubles, then quadruples, and so forth. Mount-to-mount, or bolt-eye, length is adjustable up to ½-inch to allow for better ride selection. Fournales also produces shocks for standard swing arm Harleys and they perform as efficiently as the softail versions.

Arlen Ness, as part of his Ness-Tech line, offers excellent replacement shocks by Progressive Suspension and Koni. The Konis feature four-position adjustable damping and a three-position, triple-rated, progressively wound, spring. The chrome-plated Konis are designed specifically for 1982 and later FXRs and 1979 and later Sportsters. Ness' Progressive Suspension Magnum shocks feature a hi-performance gas cell with double-wall construction and six-stage damping. These are available for FXRs, FXRDs, FXSs, and a host of Harley bikes from 1952 on, including Sportsters. Ness-Tech billet shocks, by Works Performance, offer the ultimate in style and performance. Some are offered in shorter versions to lower the rear-end of the motorcycle. The Ness-Tech billet shocks are marketed for 1979 and later FXRs and XLs, 1973 to 1986 Big-Twins, 1991 and later Dyna-Glides, and Arlen's own custom Ness 5-speed frame (Fig. 1-11).

1-11
Arlen Ness offers many rear suspension components in his catalog. From left to right: Ness-Tech billet shocks, Progressive Suspension Shocks, and Koni shocks for most swing arm Harleys.

The length of the swing arm also affects handling and is a key to the right chassis setup. Carl's Speed Shop manufactures extended swing arm kits for 1986 and later Evolution XLs. The swing arm is 3-inches over stock, enabling the street rider to "launch" hard. This swing arm provides better stability and handling by adding to the wheelbase of the stock XL (Fig. 1-12).

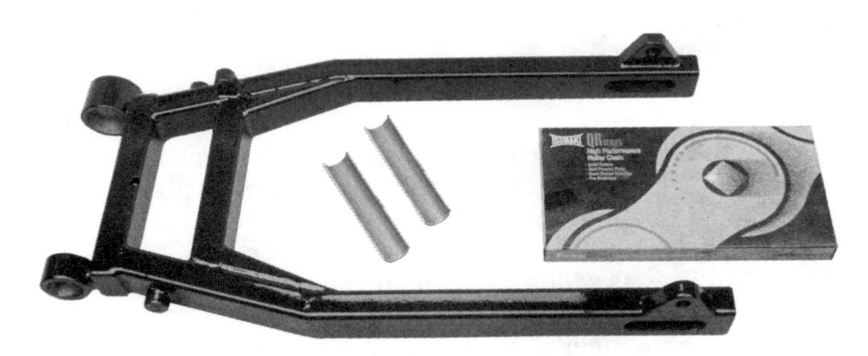

1-12
Carls' Speed Shop offers extended swing arm kits for Sportster Evolution XLs.

Aftermarket wheels

The big question with aftermarket wheels is: do they improve handling? The answer is yes, but they are an expensive, if necessary, alternative. Lightweight wheels reduce unsprung weight and gyroscopic effects and require less energy to start, stop, turn, and accelerate. This weight reduction allows the motor to spin faster, the brakes to grab harder, and the steering to be more responsive. Suspension will also benefit, with better control over bumps. Improved traction and directional control will also be realized. The newer lightweight alloys are either aluminum or magnesium. Magnesium is lighter, but aluminum is more weather-resistant and holds up better under street conditions. The hot providers of the ultra-light wheels are: Performance Machine, Marchesini, Marvic, Hypertek, and Technomagnesio. Custom Chrome Inc. offers their Rev-Tech line of forged, polished-aluminum wheels for Harleys from 1978 to the present (Fig. 1-13).

1-13
Wheels from Custom Chrome include from left to right: Revlite, Rev-Pro, Rev-Star, and Rev-Star Directional.

Some folks would like to use rear wheels up to 6½-inches wide on their FXSTs, but have been unable to because of fitment problems. Again, master bike-fabricator Arlen Ness comes to the rescue, this time with his Big-Wheel conversion kit (Fig. 1-14). Tires as wide as 100 millimeters have been used with this effective kit, which includes an offset transmission plate, a billet aluminum inner primary spacer, a motor and rear sprocket spacer, inner primary O-rings, and all necessary mounting hardware (Fig. 1-15).

1-14
Arlen Ness' Big-Wheel adaptor kit allows the mounting of 6½-inch wide tires.

1-15
Performance Machine markets Mitchell spun-aluminum wheels, which are ultra-light as well as racy looking. Wheels bolt on in the same manner as stock.

CHAPTER 2

Bottom end modifications

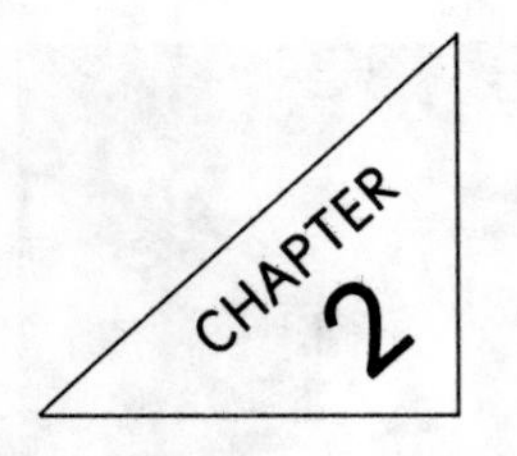

AS YOU WILL SEE in Chapter 3, the most performance gains take place in the upper, or top end, of the Harley engine. However, we can also augment the overall power of the engine by certain changes in the bottom end: stroking, camming, and so on. The "beefing up" of the lower end of the power train is accomplished by accepted methods and with the installation of aftermarket performance components, such as cams, rods, and flywheels. Perusing the following chapters will give you an insight into power train components in the upper and lower ends with much attention given to the Evolution Big-Twins, which will accept radical bore and stroke modifications.

Similar modifications apply to contemporary Sportster engines as well. The Sportster XLH-1200 is one of the fastest Harleys and possesses great potential for making it a faster and quicker machine than contemporary Big-Twins. In stock configuration, the XLH is about 100 pounds lighter than a stock 1340 cc Big-Twin and, by using lighter custom frames, can see an even greater weight reduction. The Sportster's unitized transmission and engine cases create a lighter, more compact, easier aligning powerplant than the separate engine and tranny configuration, which is prone to power transfer loss.

The XL features Harley's hottest camshafts, four of them, making the engine more responsive to improvements in ignition, intake headwork, and exhaust systems, without a camshaft change. Sportster engine design is also very acceptable of increases in cubic inches. The bigger XLs can be punched-out up to .498 inch for added horsepower gains. XL engines respond well to modification and can handle more power than that produced by the XL engine in stock form (Fig. 2-1).

A higher lift cam set will augment valve train performance. Andrews markets an excellent line of XL Evo cams, in four grinds, that can be utilized with stock pushrods and valve springs. Though not mandatory, it is wise to include high-lift valve springs with high-lift cams, since they are lighter, stronger, and harmonically wound. They provide better clearance and, due to their resilient strength prevent valve float at high rpm's.

2-1
The Evolution Sportster engine is a new design upgraded version of the potent sportster mill of yesteryear. It will yield additional power with today's aftermarket performance parts.

Stroker engines

A stroker engine's displacement is increased by replacing the stock flywheels and rods with an aftermarket crank assembly having a longer stroke than the standard Big-Twin or Sportster. The stroke is the distance the piston travels from the uppermost position of its movement, or Top Dead Center, to the lowest extreme of its travel, Bottom Dead Center. The distance traveled by the piston is a result of centerline spacing between two centrally located shafts in the flywheels. On the left side of the crankshaft is the motor sprocket shaft (Big-Twins) and on the right side is the pinion shaft. Stroker flywheels increase the distance in order to obtain an increase in piston travel distance.

Strokers can be as reliable as any performance engine provided they are properly and painstakingly assembled with quality components and periodically checked and maintained. Stroking is an operation performed on the lower end of an engine, although it may lead to minor or major changes in the top end.

Methods of stroking an engine

To increase the crankshaft offset, a stroker lower end has the crank pin, the bearing journals that the rods ride on, moved farther away from the centerline of the crankshaft (Fig 2-2). Some cranks are built-up with individual mainshafts, flywheels, and crank pins pressed together: Harley, Vincent, and most European

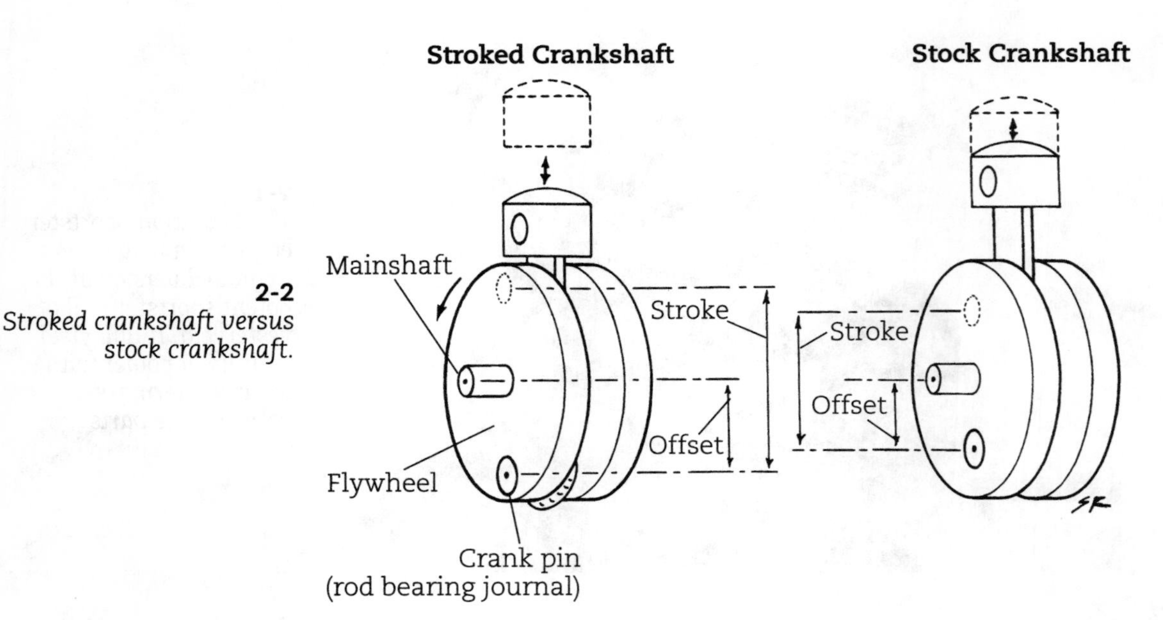

2-2
Stroked crankshaft versus stock crankshaft.

and Japanese engines. When new or reworked flywheels with crank pin holes located "further out," away from the center mainshaft are used, we have the basis for a stroker bottom end.

The distance the offset is moved outward will augment the stroke in two directions, first as the piston rises to the top of the cylinder, and again as it falls to Bottom Dead Center. Stroke is actually twice the total offset from the crankshaft's centerline.

Compensating for longer stroke

With the pistons traveling into all sorts of new territory, problems can arise if you just drop a stroked crank into an otherwise stock motor. Because the piston domes are rising higher on each revolution, you will get at least a sudden, extreme, and probably undesirable boost in compression ratio. More likely, your buckets will push out so far that the tops will conflict with the valves and might run into the cylinder head with a clang. On the lower half of the revolution, stock pistons might kiss the top faces of the flywheels or main bearing retainers in a very unfriendly manner.

Fortunately, there are ways to make allowances for the longer piston travel of a stroker. Most "bolt-in" stroker lower ends include special rods, shorter than standard, or have a lowered wrist pin hole that keeps the pistons from riding up too far. Some stroker flywheels are of smaller diameter than stock to prevent bottom fouling.

Another alternative is the use of special "stroker" pistons. These have a shorter skirt to clear the flywheel tops, and less dome and height above the wrist pin to compensate for the longer travel. Because of their very short design, some stroker pistons include Teflon-type skirt buttons to control rocking and prevent *galling*.

Longer stroke can also be accommodated by using longer than stock cylinders that move the cylinder heads back to their original positions relative to the piston tops. Stroker cylinders might necessitate the use of longer pushrods and oil lines.

Finally, compression or stroker plates can be placed underneath stock barrels to achieve the same result, raising the cylinder a distance equal to the thickness of the plate.

On moderately stroked engines, one or two of the above compensating methods might be enough to put everything back in place. Very long strokers will need more. They may pull the ends of the rods out of the cylinders at such an angle that notching and relieving of the cylinder walls, crankcases, and the rods themselves might be necessary.

Strokers—Guidelines for particular engines

Milwaukee iron is particularly suitable for stroking because very large bore sizes are limited by the close location of the tappets and followers to the cylinder bores. There is a good aftermarket supply of Harley stroker components, particularly from S&S and Axtel.

S&S offers complete kits to turn your V-Twins into 89- or 96-inch strokers. The 89-cubic-inch kit requires no special machining. The kit includes flywheels and mainshafts, heavy-duty rods, cast pistons, S&S Super E carb, S&S cam, high-performance valve springs, and adjustable steel pushrods with hydraulic lifter travel limiters.

The S&S 96-cubic-inch set-up turns your Evo into a 96-cubic-inch "beast of the streets." This kit offers power and torque to spare. Featured are the famous S&S Sidewinder 3⅝-inch big bore cylinders and pistons and an S&S 4⅝-inch stroker flywheel and rod assembly, with sprocket and pinion shafts. Also included are: S&S cam, valve springs and collars, S&S adjustable pushrods and hydraulic lifter travel limiter kit and the fabled S&S Super E carburetor (Fig. 2-3).

2-3
S&S offers some potent stroker kits. This is their 96-cubic-inch kit for the ultimate big bore and stroke.

This book recommends new, ready-made stroker flywheels for Harleys. These are designed to fit into a stock engine without a lot of finagling, are much stronger than stock items, and no rebalancing is necessary. Much larger increases in engine size are available, up to about 70 cubic inches for a stock bore Sportster and 92 cubic inches for a Big-Twin.

Stroker pistons are available from MC and Axtell, longer cylinders from Axtell, and stroker plates from several sources. Because the Harley engine design is a V-twin, a couple areas require particular attention. Because the cylinders are at angles to each other, longer barrels position the intake ports farther away from each other. This increased distance makes a longer carb manifold necessary and, luckily, S&S makes one for various stroke-length engines. Also, since Harley rods run one inside the other, some relieving might be necessary for very long strokers. In addition, since the size of the rod forgings vary, it is hard to set down any absolute guidelines.

With moderate strokers, stock Harley cylinders can be used with compression plates and stroker pistons. On 74s, if the stroke is 4¾ inches or more, the best procedure is to drill the oil return hole straight down into the crankcases and close off the old one. Seventy-four-inch stroker cylinders from Axtell already have the notch and the low oil hole. If you plan to run a big bore set-up, along with your stroker, you'll probably be opening up your crankcases enough anyway that crankcase cutting won't be necessary. However, the inner edge of the piston skirts will hit at BDC unless trimmed slightly.

Beyond that, the only other parts you'll need are extra length oil lines, pushrods, covers, etc., and most of this is available from S&S in their complete stroker kits. Also, if you go to a real long arm and super tall cylinders, the height of your engine will probably require a reworked top motor mount, journals built up with welding and then reground as far out as possible.

Strokers—A *few* conclusions

With any stroked engine, complementary speed equipment should be installed to maximize the power benefit of the extra added inches. Improved cylinder and crankcase breathing will help by means of enlarged breather passages, wider ports, and bigger valves. Hotter cams, specifically designed for strokers, not only boost horsepower, but the increased overlap reduces compression pressure at kick-over speeds and makes starting a stroker less of a pain. In addition, stroked engines usually fire up and run better with ignition timing retarded one degree or more from stock specs.

Although the temptation is there to go all the way, remember that moderately stroked engines are easier to build and maintain on the street. The longer the stroke, the more clearance problems you'll have and the higher piston speeds will be.

Finally, even if you're a pretty fair mechanic, it's a good idea to enlist some local help to build a stroker that does not fall into the bolt-together category. Most every part of the country is bound to have a hard-core engine builder who has some experience in the field. Most drag-race engine builders are well versed in this area, since stroked engines perform well at the drags (Figs. 2-4 and 2-5).

2-4
Stroker and flywheel big bore kits are available from S&S for older 74s and Sportsters. Kits up to 92 cubic inches are available.

2-5
S&S iron head Sportster stroker kits up to 4⅝ inches are available for punching out older Sportster mills. The rods and shafts are optional.

A most highly recommended engine builder, who has been stroking and building record holding Big-Twins for decades, is Carl Morrow of Carl's Speed Shop, Santa Fe Springs, California. Carl's shop has the machining facilities to take on all types of engine work, including boring and stroking.

Truing flywheels

Everyone knows Harley engines vibrate. The amount of vibration can be minimized by perfecting the running mass of machinery in the crankcase lower end. The vibrations originate from a number of sources. Two we can most easily contend with and improve upon are unbalanced reciprocating engine mass and untrue flywheels.

Harley twins have cylinders in a 45-degree configuration, putting power pulses at alternate intervals, which produces the uneven Harley idle. Little can be done about the uneven Harley power pulse vibration pattern, but the lower end mass characteristics can be altered to a great degree. Unbalanced reciprocating mass vibrations can be eliminated, or greatly reduced, by sending flywheels and rod-and-piston assemblies to an engine rebuild specialist. Flywheel *truing* can be performed on a truing stand or similar fixture and flywheel *runout* checked with dial indicators. According to Harley, runout on each wheel should be no more than .001 inch. It is possible to adjust the wheels to ½ or ¾ of a thousandth runout. This will produce a great reduction in vibration.

Maintaining a “true running” condition is crucial. Flywheels are positioned by means of a wedge-type fit on an unkeyed or unindicated tapered shaft. The force of the power stroke can knock the flywheels out of alignment and it might be necessary to true flywheels whenever excessive vibration is detected. If you’re a better-than-average wrench and not timid about going into the bowels of your crankcase, you might want to true the flywheels at regular intervals to ensure against adverse vibrations. A long-lasting truing job is dependent on proper flywheel assembly. The pinion shaft and crank pin always mount to the right flywheel and the sprocket shaft to the left flywheel. Check the Harley manual for the proper procedures.

First the crank pin nut is installed, lightly tightened. A steel ruler, or similar straightedge, is held against the edges of the flywheels. The top wheel should be rotated until the straightedge rests evenly on both wheels. Done at a point approximately 90 degrees from the crank pin, this puts the wheels in approximate concentric alignment. Tighten the crank pin nut with a socket and breaker bar, just snugly. If the crank pin is set too tightly, adjustment is difficult without bearing down too hard on the flywheels. If too loose, adjustments won’t stabilize.

Truing procedures

Mount the flywheel assembly on the truing stand so the wheels turn freely but with no end play. The sprocket shaft should be on the right. Adjust dial or truing stand indicators to read out mid-range. The pickup points on the pinion and sprocket shafts should be as close to the flywheels as possible. Slowly revolve the flywheels and observe the runout indicated on the dials. Indicator deflection toward the flywheel shows a high point in the assembly and chalk marks will serve to indicate the high points. To true the wheels, tap lightly on the high points with a lead hammer or mallet, then replace the crank assembly in the truing stand and check deflection again. It will probably be necessary to repeat the procedure several times until the wheels run true.

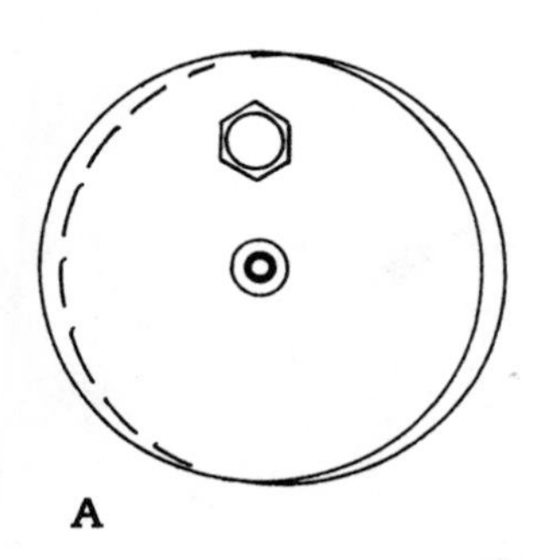

Figure 2-6 shows the three basic forms of runout. In most cases, they go out of whack in a combination of two of the three forms together. In such a case, correct each condition one at a time. *Toe-out*, or wheels that are "bowed," are too far apart in the quadrant opposite the crank pin. Tighten a large C-clamp onto the wheels at the high point and tap at the rims near the crank pin. Use light taps to begin with, increase them in small increments until the desired change is made.

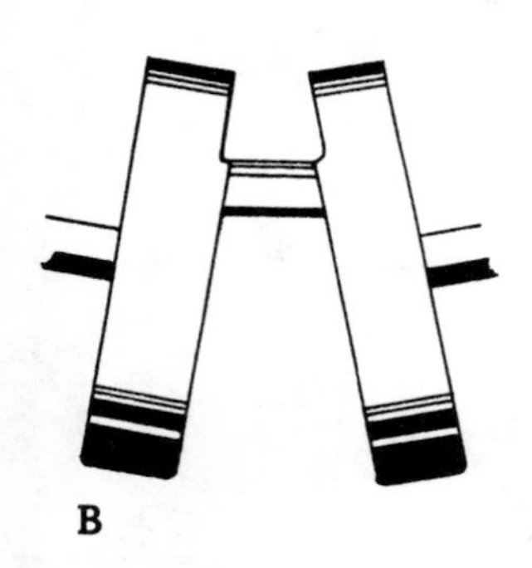

Toe-in or "pinch" is corrected by inserting a wedge between the flywheels, across from the crank pin. Overspread the wheels slightly more than they are pinched. When the wedge is removed, the wheels will spring back into place. You must always allow for "springing" when trying to alleviate pinch or bow. Truing flywheels is not necessarily a difficult job, but it is tedious and needs to be done slowly and exactingly.

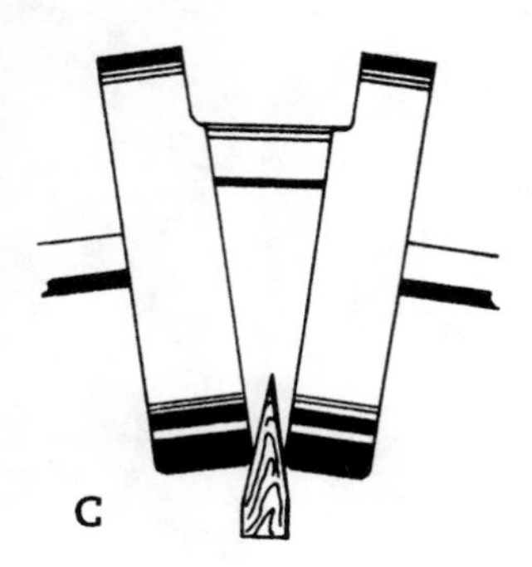

2-6
Examples of flywheel runout. A. Up and down misalignment; B. "Pinched"; and C. "Bowed".

Cases

Stock Harley *cases* are great for stock bore and stroke configurations. However, for a performance engine, you can't beat the bullet-proof billet-machined cases provided by aftermarket manufacturers. S.T.D. produces alternator-type cases for 1970 through 1983 74s and Evolutions. They also offer generator-style cases for 1955 through 1969 and 1948 through 1954 vintage Harleys. Sputhe markets similar cases for big bore and other applications. Some of the finest, and a particular favorite of street and strip riders, are the Delkron cases, American made for contemporary Evolutions and selected vintage Harleys. The Delkron cases are delicately machined from quality aluminum in standard and big bore versions (Fig. 2-7).

2-7
Special big-bore engine cases are available from Delkron for the new Evolution engines. These are a must for big bore and stroke enthusiasts who require a strong, indestructable case.

Cams

Cam efficiency and power relates to engine displacement and flow. This does not mean that power and performance are necessarily the same. Sophisticated engine refinements are irrelevant if in the final outcome they are inhibited by incorrectly timed cams. A cam that performs well over 100 miles per hour may be a disappointment at 50. It is a common fallacy that the more radical a cam is in terms of lift and duration, the more power realized. A cam that causes an engine to stutter at idle does not necessarily make it a terror at 5,000 rpm (Fig. 2-8).

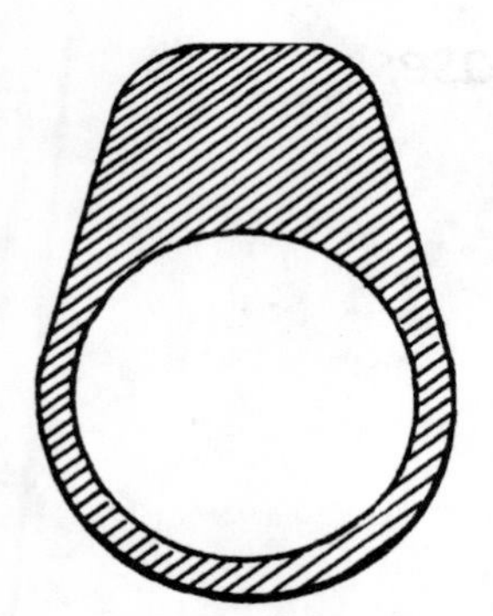
Tangential

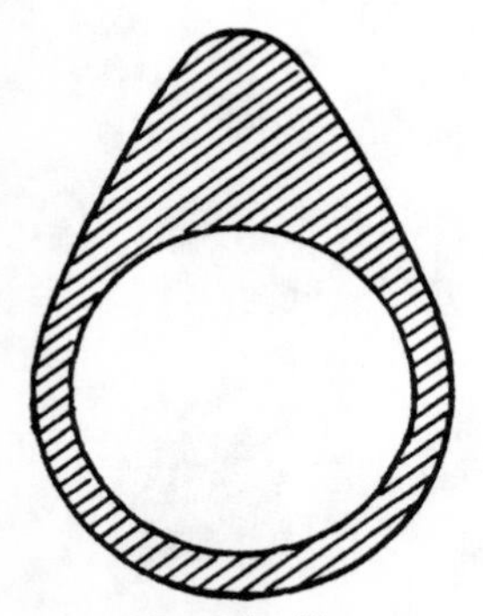
Parabolic

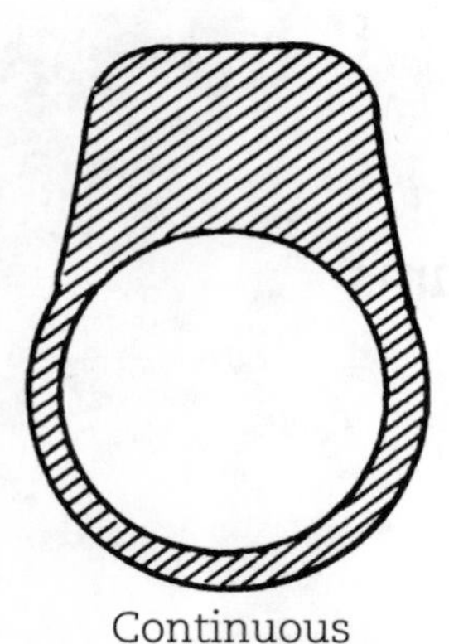
Continuous acceleration

2-8 *Cam lobe types.*

The cam's primary function is to convert rotary motion to linear, or up-and-down motion. The cam controls the amount the valves are lifted from their seats, in a specific pattern, timing the opening and closing of the valves relative to crankshaft rotation.

Cams, in general, fall into three shape categories, or variations of three basic themes. The three types are the tangential, parabolic, and constant acceleration. The tangential employs flanks, which are tangent to the wheel or base diameter of the cam. Tangent lobes may be parallel as with the Harley Knucklehead, or slope in toward each other at a specific angle to produce the desired valve lift characteristics. Valve motion relative to the tangential cam is on nontransitional acceleration. As the cam follower travels off the heel profile, it suddenly becomes engaged in quick acceleration. There is no starting ramp in the profile to allow a gradual buildup of motion.

The parabolic profile cam, or "mushroom type," is identified by its convex parabolic profile. The advantage in this type is that it provides constant movement of the valve at the onset of the opening stroke. With this profile, most of the valve lift is placed at the onset of the stroke, or opening motion, and also in the lift on the closing stroke. The result is that the valve is open during the greatest lapse of cam duration. This is beneficial in terms of engine breathing, but can impose limitations as to how much valve overlap can be integrated into the lobe pattern without causing opposing effects. The parabolic profile creates a gradual slow down of the valve components at the termination of the opening stroke, plus a similar acceleration characteristic on the closing stroke.

The cam shape most popular for performance is the constant acceleration type. The shape is a concave parabolic contour allowing long acceleration and deceleration times at the end of the strokes, keeping rate increases even. This minimizes the stress imposed by valve train acceleration and opposing effects of harmonic valve spring compression.

Cam contouring is based on cam timing relative to piston location in a four-stroke cycle. This, plus calculated overlap of intake and exhaust durations, helps govern the physical shape of the cam itself.

The air/fuel mixture does not begin to move at the initial opening of the valve because inertia has to be overcome. The cam designer must consider how much in advance the valve must open to begin to move the fuel mixture. Also determined is how much overlap of valve openings is necessary not to waste fuel mixture in an effort to clear out exhaust gases. Inertia, delaying the flow of fuel mixture into the combustion chamber, acts on the flow after the piston reaches bottom dead center so that some air will still flow in after the piston begins its next compression stroke. For optimum breathing, the cam must initiate closing as soon as the upward travel of the piston retards the airflow, prior to reversing direction, forcing the intake mixture back into the intake manifold. Cam shape can only be calculated on a narrow band of engine rpm. Below the desired rpm band, the valve will remain open too long, forcing a part of the fuel mixture backwards and creating the rough idling and raspy exhaust characteristic of a racing or performance engine. Above that calculated rpm band, engine breathing will be erratic and less efficient. The objective of the cam designer is to balance all the factors that contribute to engine efficiency while bolstering power at the speed most often utilized by the given engine. Cams are referred to in performance circles as "mild" to "wild." A "mild" cam is one designed to promote performance at low to medium engine speeds. A "wild" cam will do the same, but will do it better at higher engine rpm's. Properly formulated, each type of cam is most efficient in its proper speed range (Figs. 2-9, 2-10, and 2-11).

The valve train, working in conjunction with the cam, must also be considered. High-lift cams may cause stock valve springs to *coil bind*, the condition in which springs are compressed until the coils contact. *Spring surge*, another detrimental condition, is most common on the expansion stroke when different spring sections do not expand or contract at uniform rates. These conditions can cause high-speed rebounds, which cause an expansion faster than that of the overall spring, which then re-creates random compression

2-9
Crane markets their standard "Fireball" series cams of their "Hi-Roller" series, with adjustable advance and retard timing settings.

2-10
Andrews markets a full line of high quality cams for Evo 1340 engines and Evo X-4 Sportsters. Andrews also provides special grinds for virtually all applications. Right and left, XL cams. Center, Evo 1340 cam.

within an expanding spring. Spring surge can be caused by lobe designs that accelerate the valve train unevenly or too rapidly. It may manifest itself when springs are too weak or tension too light for the corresponding valve train, adversely affecting the acceleration and deceleration of the valve train.

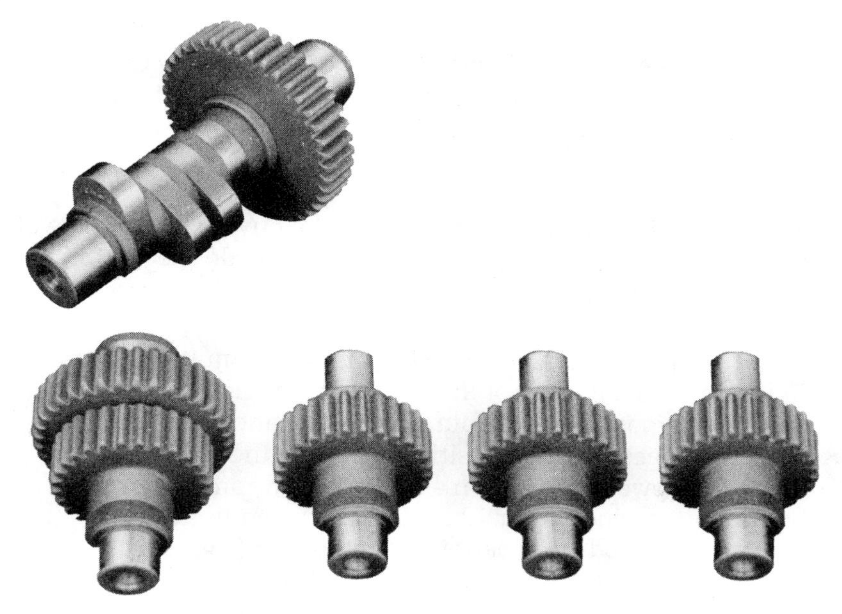

2-11
Carl Morrow offers special grind performance cams for both Evo Big-Twins and Evo Sportsters. Left: Evo; Below: Evo XL.

Valve float, another undesirable response, occurs when the value does not respond to the tension conditions of the existing valve train, allowing the valves to "float," or remain hanging open over their seats. This condition is serious, since the valves can hit the pistons, causing severe engine damage.

Another negative condition is *valve bounce*, caused by cam lobes without gradual deceleration ramps. This allows the valve to overshoot its stroke upon opening and causes it to bounce off the seat when closing. The solution requires the selection of a lobe shape incorporating clearance ramps and gradual, uniform, acceleration rates. A clearance ramp is contoured into the lobe at the start of duration, taking up clearance in the valve components before the valve is accelerated to full open position, with a similar ramp at the final cam duration point redistributing train clearances.

2-12 *Crane "Hi-Roller" series cams offer multi indexing. One position gives four degrees of advance and the other retards the original timing four degrees. Overall, an eight degree timing range is available with the "Hi-Roller" series.*

High tensile strength springs are a must with performance cams. These springs and spring kits are marketed by Crane, Sifton, S&S, Rivera Engineering, and many other performance specialists who provide performance cams.

All cam manufacturers produce high-quality cams slightly differing in specifications. This allows the buyer a wide range of performance choices so the cam selected will do the job for which it is intended.

Crane offers adjustable-gear cams that allow multi-indexing. This allows installation of the cam in the original straight up zero degree position ("0"), four degrees advanced position ("A"), or four degrees retarded position ("R"). Advancing the cam drops the power range down a few hundred rpm's, creating better bottom end power. Retarding the cam raises up the power band, producing better top end power (Fig. 2-12).

Installing a cam

The accompanying illustrations show the tried-and-true procedure. A Harley shop manual will greatly assist in disassembly. After disconnecting the battery, remove the exhaust, footrest, brake pedal, and air cleaner to allow access to the pushrod covers and gear case. Remove the pushrods, lifter covers, and spark plugs. Turn the engine by rotating the rear wheel. The engine must be in gear to do this. Turn the engine until both front cylinder valves are closed. You will see both front cylinder tappets in the down position and the piston at the top through the spark plug hole. Loosen the pushrod adjuster locknuts and remove the pushrods from both front and rear cylinders. Keep the pushrods aside in their respective order for reassembly. Remove the ignition cover, bolt, and rotor, or point ignition, if your model Harley has one. Next, remove the gearcase cover and gasket to access the cam (Figs. 2-13 and 2-14).

Remove the camshaft assembly, spacing washers, and thrust washers. Do not discard the spacing and thrust washers as they must be reinstalled with the new performance camshaft.

Please note that when using high lift cams it is necessary that some minimal machining be done on the crankcase to prevent

2-13
Carefully dislodging the gear case, with a screwdriver wedged behind it, also moves the stock cam forward for easy removal.

the high lift lobe from hitting the crankcase wall (Fig. 2-15). Insert the new cam and align the timing marks as shown in Fig. 2-16.

The final step is the reinstallation of the cam washers or shims and the replacement of the gearcase cover, tightening it to the proper torque. It is recommended that, following the cam change procedure, a dial indicator is used to check side play of the cam shafts (Fig. 2-17).

If excessive side play is noted, you might have to add shims to take up the excess. Replace all the engine components initially removed and you're ready to roll with a more potent engine. The installation on the Evo engine, shown here, was performed by master mechanic and "Monster Motor Builder" John Sachs, of Ft. Lauderdale, Florida.

2-14
Removing the ignition mechanism and ignition cone bares the stock cam, which can be easily slipped out.

2-15
The pencil point shows the crankcase area around the cam ground down to prevent higher lift cam lobes from hitting the crankcase wall.

2-16
The new cam is inserted in place of the old, making sure that the aligning marks on the gears match up. The high lift cam installed here is an Andrews EV-5.

2-17
Use a dial indicator on the camshaft to check the new cam's side play. If side play is excessive, install shims to take up the play.

Important notes (1948 through 1969 engines)

Harley engines manufactured from 1948 through 1957 utilized a bushing in the crankcase to position the camshaft. 1958 and later engines make use of a needle bearing. Many current performance cams are manufactured for the needle bearing type. If your engine contains the straight crankcase bushings, it must be converted to needle bearings by a qualified dealer.

Early model engines, 1948 through 1969, use a series of gears to drive the ignition system. These engines require a short nose, ¾-inch-long, camshaft. The "circuit breaker" gear is closely positioned to the front lobe of the cam and contact between the gear and high-lift cam lobe is possible. Clearance between the lobe and circuit breaker gear must be carefully checked to ensure that clearance is at least .030 inch. If not, the gear must be modified by machining away some of the area where the lobe contacts, keeping enough gear width to drive the ignition system. Later, 1970 and on, Harley engines drive the ignition off the end of the camshaft with a long nose, approximately 15⁄16-inch, camshaft.

Between 1948 and 1984, Harley utilized two different-style cam gears. The two, as well as the matching pinion and breather gears, are not interchangeable. Engines from 1948 through early 1977 have a "straight cut tooth" cam gear. This gear has no distinguishing identification marks. Its matching pinion gear has teeth across the full width of the pinion gear body. The matching breather gear has a slash mark to identify its timing position.

Late 1977 to 1984 (non-Evolution) engines make use of a cam gear with curved teeth, identified by a groove cut into the gear face. The matching pinion gear has teeth across half the width of the pinion gear body. The matching breather gear has a cross to identify its timing position.

Setting gear lash

To establish proper gear *lash*, cam gears and pinion gears are made in different pitch diameters. The different sizes are indicated by color codes on both gears. Check the color code chart in your Harley-Davidson manual for the exact sizes. It is advisable to match the color code of the cam gear, if available, to the color code of the original pinion gear. Crane offers three cam gear sizes in .001-inch increments in pitch diameter. They are coded yellow, red, and blue for small, medium, and large. If the gears are not properly matched you will hear some gear noise. If the gears are too loose you will hear "clatter," much like a lifter noise. If the gears are too tight, you will bear a gear "whine." When properly fitted, you will hear a very slight gear whine when the engine is hot. You can also fine-tune gear lash by changing the stock pinion gear to the next larger or smaller size, as necessary.

Prior to running your bike with an upgraded cam set-up, check valve clearances to make sure they don't hit on the overlap periods. Also, check valve to piston clearances. Your Harley shop manual will tell you how, or your dealer can do it for you. For Evolution Sportsters, you can't beat Andrews' wide selection of the finest Sportster cams. All Andrews cams will run to 6500 rpm with stock hydraulic lifters. Andrews E-V Sportster cams have stock base circle sizes so that stock, nonadjustable, pushrods can be used, except V9 or BV. Adjustable aluminum and chrome moly steel pushrods are marketed by Andrews for use with their cams (Table 2-1).

Table 2-1 Touring and high-performance cams

Part#	Year	Grind	Timing(*)	Duration .053	Duration .020	Max Lift	TDC Lift	Application
Stock	(88-91)	**D**	02/41	223	270	.458	.094	Listed for comparison.
		D	41/02	223	270	.458	.094	(Note: 1985-1987 exhaust cam lift is .414)
298120	(86-90)	**V2**	22/38	240	290	.465	.180	Bolt in cams for stock 883, 1100 or 1200 engines. More duration and lift means extra power thru RPM range. Stock springs and hydraulic lifters recommended. 2000-6000 RPM.
298125	(91-up)	**N2**	46/18	244	290	.440	.155	
298140	(86-90)	**V4**	30/46	256	296	.490	.216	Street/drags: Stock or modified 883/1100/1200. Slightly higher idle speed but stock springs-hydraulic lifters recommended. RPM range: 2500-6500.
298145	(91-up)	**N4**	52/24	256	296	.490	.189	
298160	(86-90)	**V6**	34/50	264	302	.490	.241	Modified engines with stroker flywheels. Stock springs and hydraulic lifters recommended. RPM range: 2500-6500.
298165	(91-up)	**N6**	56/28	264	302	.490	.212	
298180	(86-90)	**V8**	32/44	256	296	.490	.226	Modified 1100-1200, stroked 883's with stock springs and hydraulic lifters. Great mid-range power: 2000-6000 RPM.
298185	(91-up)	**N8**	56/28	264	302	.490	.212	
The following two grinds are Andrews Iron Sportster cam shapes with the correct timing for the different roller diameter and rocker ratio. They are proven cam designs and for larger EV Sportster engines will produce lots of smooth usable power.								
214210	(86-90)	**V9**	33/53	266	309	.555	.240	Med. lift cams for stroked engines from 80-88 inches. Broad torque range to 6000+ RPM with hydraulics. Andrews springs, collars and pushrods required. (1991-up need long pushrods).
214215	(91-up)	**N9**	53/33	266	309	.555	.240	
214265	(86-90)	**BV**	35/59	274	316	.590	.260	Hi-lift cams for 88+ inches. Andrews pushrods, springs and collars required. BV cams start easy and run strong; 2000-6000+ RPM with hyd. lifters. (1991-up need long pushrods.)
214268	(91-up)	**NV**	59/35	274	316	.590	.260	

(*) Timing listed for .053 lift figures.

Note: '91-up Sportster engines are considerably different than earlier EV Sportsters! Cam gears for earlier EV Sportsters will not fit '91-up engines. For that reason new part numbers are listed for '91-up engines.

Andrews also provides cams and cam gear kits for all the older iron head Sportster engines, in various lifts and durations.

Tips on choosing a cam

Performance buffs choose modified cams for the express purpose of power gain. Overcamming can sometimes result in power loss, opposed to what is initially desired. While higher lift cams produce more peak output, at high rpm's, a milder cam may be more applicable for the street, where some low rpm power is important.

Larger engines can apply more radical cams than smaller mills. For example, an 88-cubic-inch Harley engine can handle a hotter cam than a stock 74 engine. The added cubic inches can make use of the greater airflow resulting from a cam with higher lift and longer duration. Another important factor involved in selecting a cam that applies especially to Harley engines is compression ratio and fuel detonation, commonly known as "pinging." Lowering the compression ratio of an engine results in less power at all engine speeds. A reduction in pinging can be achieved, but at the cost of reduced power output. A solution for the pinging syndrome involves a cam swap leaving the compression ratio at about 8.5 to 1. A longer duration cam has the same effect on low speed compression pressure as a lower compression ratio. The reduction of compression pressure at low speed will result in the elimination or reduction of pinging. With a longer duration cam and higher compression ratio factor, power in the mid and high range will be increased. This fact, relative to performance cams, has not been closely studied, but it is a given fact. Many of the newest American automobile engines have compression ratios of 8.5 to 1, or higher, and they burn lower octane fuels without exhibiting pinging.

Reducing compression ratio lowers the degree of "squeeze" on fuel being compressed into the cylinder. The pressure reduction results in less compression generated heat. With the lessening of heat the fuel has less tendency to "ping" prior to spark plug firing. With a long duration cam, compression pressure in turn is reduced, since actual closing of the intake valve occurs with the piston at a higher level in the cylinder. The piston, at this point, has a shorter distance to travel to the spark plug firing location.

Less distance to the firing location and less distance on the compression stroke translates into a lower compression ratio and, no pinging. This effect is also realized with 8.5 to 1 pistons resulting in no pinging in lower speed ranges, where pinging occurs, and with more power in the mid and high rpm range.

Breather valves

After your cam change, you might want to replace your stock Harley plastic breather valve with the more efficient steel units marketed by Custom Chrome or S&S. The Harley plastic fiber material is abrasive and tends to score the right case area. The stock material is not as strong as metal and can become out-of-round. An out-of-round breather valve can result in vacuum seal breakdown and cause problems such as smoking and excessive oil seepage from the breather vent hose. These problems can be alleviated by simply replacing the stock piece with a Custom Chrome or S&S breather valve.

2-18 *The stock Harley plastic breather valve and, at the left, the more efficient Custom Chrome unit. A similar steel replacement valve is available from S&S.*

Correct breather timing is essential to good engine operation, allowing the crankcases to ventilate properly. As you increase cubic inches, you must adjust the breather valve to stay open longer to minimize crankcase pressure, which builds up because of the increased volume that must now be displaced. The metal valves have additional timing built into the valve. Another feature of the replacement valves are removable gears which allow dialing-in of the valves for individual applications (Fig. 2-18).

Clutches

Those looking for upgraded clutch performance may avail themselves of the new Kevlar kits from such companies as Barnett, long known for their quality clutch plates and assemblies. Asbestos was the old choice for reducing heat and producing friction in clutch assemblies. However, since the material has been found to be carcinogenic, Kevlar, a space-age material, has taken its place and proven to be trouble-free and more effective than asbestos.

For all-out state-of-the-art clutches and clutch assemblies you can't beat the Bandit Super Clutch. These clutches are available for both Big-Twins and Sportster XLs as well as Shovelhead and iron head XC Harleys. The Bandit clutch provides twice the lining area on each plate, which is Barnett's Kevlar material. This accounts for the positive action of the Bandit kits, which include friction plates, springs, spring cups, spring collars, hub, pressure plate, and lock nuts to retain the springs (Figs. 2-19, 2-20, and 2-21).

2-19
American made Bandit Kevlar clutches for Big-Twins and X-Ls feature double the lining area on the plates. High-quality billet aluminum is used for hubs and covers.

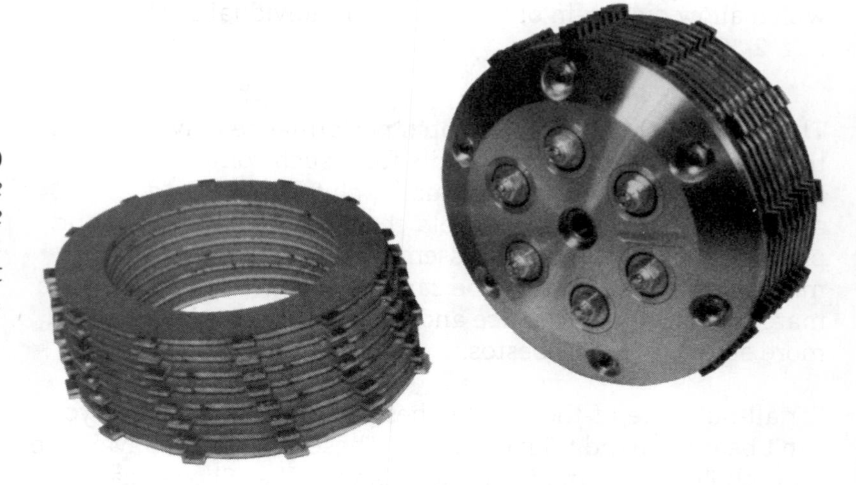

2-20
Bandit Super Clutches feature a wide lining area on the plates and a beefier housing. These are the strongest, most efficient, clutches available.

2-21
Housing cover plate and drum are heavy duty with quality machining, the best clutches available for Evo Big-Twins, Sportsters, and older model Harleys.

Transmissions

"Trannys" are trannys and, as Harley makes them, most efficient. However, there are a couple companies that offer special tranny packages with smoother shifting replacement gears, made by companies such as Andrews (Figs. 2-22 and 2-23).

2-22
Sputhe offers a tranny conversion kit, 4- to 5-speed, that will mount in a 4-speed frame and includes all the necessary components.

2-23
Assembled Sputhe tranny case ready for insertion in frame. Casemates up with stock Harley fittings and covers.

Sputhe, along with Andrews, offers a 5-speed tranny and gear package, which allows a four- to five-speed conversion for pre-1980 4-speed frames. The Andrews kit provides a close ratio first gear and replacement gears for the other four. Lead-in ramps have been machined into the gears to improve shifting and reduce wear on drive *dogs* and drive slots. Assembly instructions are included by Sputhe and Andrews. The Harley manual will also indicate proper installation procedures.

A wide array of Sportster main drive sets and 5-speed tranny alternatives are also marketed by Andrews, for contemporary and older Harley models. I strongly recommend the new Andrews catalog to everyone. It is chock-full of authoritative data and information on their full line of cams, gears, and performance components (Figs. 2-24 and 2-25).

2-24
Andrews sells a 2.94 close-ratio first gear-set, left and right, providing close ratio shifting into second and 5 extra mph at peak rpm. At center is the stock main gear for belt drives, also sold by Andrews.

2-25
Andrews offers 5-speed replacement gears for the Evo XL 5-speed.

Belt drive

A number of newer Harleys utilize belt drives as an OEM feature. A few companies offer performance belt-drive units that assist in power transfer through state-of-the-art pulley and belt systems. Conversion kits are also marketed that will convert older Harleys to belt drive, with minor modifications. Karata markets conversions as does Rivera with their "Primo" line (Fig. 2-26). The Rivera "Primo," 11-millimeter primary belt drive and pulley kit, has a combination of 31 teeth on the front pulley and 47 teeth on the rear. With more tooth to pulley contact, a greater *wrap* is created for the drive belt, which helps extend belt life. An added benefit is the reduced power to tooth ratio. The 11-millimeter "Brute II" drive kit, complete with idler bearing and featuring a Gates reinforced rubber belt, is available in six versions.

2-26
Sandy Roca's Shovelhead street chopper features Karata belt drive and a see-through primary case cover for a novel effect.

Rivera also offers a full line of Harley-Davidson 74 and 80 belt drive kits for closed or open primary applications. The open primary kits are intended primarily for racing purposes but can be used on street machines. Closed primaries offer more belt and pulley protection from dirt, debris, and stones. The 74 & 80 kits are 8-millimeter drive kits and are available for a number of 1937 and on 74s and Shovelheads. Pulleys are hard anodized and all front pulleys are drilled and tapped for easy removal. Models with one-piece aluminum pulleys have hardened clutch dogs, secured by three steel rivets (Fig. 2-27).

2-27
Larry Stanley's Shovelhead dragster features "Primo" belt drive in an open-primary set-up. The extra-wide belt offers optimum engine power transference.

Lifters & pushrods

There is some debate as to the benefits of solid lifters versus hydraulics and vice versa, mostly concerning Big-Twins. Sportsters never used hydraulic lifters prior to 1985. Each type of lifter has its own advantages.

In a Big-Twin, most Harley camshafts will function properly with hydraulic lifters if the engine is set up correctly. Hydraulic lifters provide quiet operation, less servicing, and no loss of lift and duration due to heat expansion. If, at installation, the oil holes in the lifter blocks are in correct position to feed the lifters, when the cam is positioned at the lowest left-hand point, the lifters will operate properly. It might be necessary to groove the lifter blocks for better oil flow to the feed hole on the tappet body.

Solid lifters exhibit other advantages. They are easier to install and set and not as affected by dirty or gummy oil as are hydraulics. On high-performance dragstrip and racing machines, solid lifters may be the best choice; however, for everyday street riding, low-maintenance hydraulic lifters are a better alternative. The newer Evo hydraulics are so refined that unless you encroach into the 6500 rpm range you should stick with them. Shovelhead hydraulic lifters are not as sophisticated and a little harder to work with, but the hydraulics make for a quieter engine that requires less servicing.

As for pushrods, the trend is toward the adjustable type, which require less attention and provide quick adjustability. If you favor solid pushrods, you might want to consider the Crane Shovelhead adjustable-mechanical roller tappet and pushrod set. The 3-1005 kit is for stock length cylinders and the 3-1006 kit is for the long stroke cylinders. These kits are specifically designed for 1966 to 1984 Shovelheads.

The kits eliminate the need for adjustable pushrods, since the lifter units adjust. The roller wheel is supported by needle bearings, which assist in reducing friction. Threaded adjusters allow you to adjust valve lash at the tappet (Fig. 2-28).

2-28
For use with solid pushrods, Crane markets their lifter-pushrod sets with adjustable lifter tappets.

Crane also offers mechanical lifter and pushrod sets for older Shovelheads. These are 33-percent lighter than stock components. These mechanical tappets have altered pushrod seat heights, resulting in less weight and reduced guide block wear (Fig. 2-29).

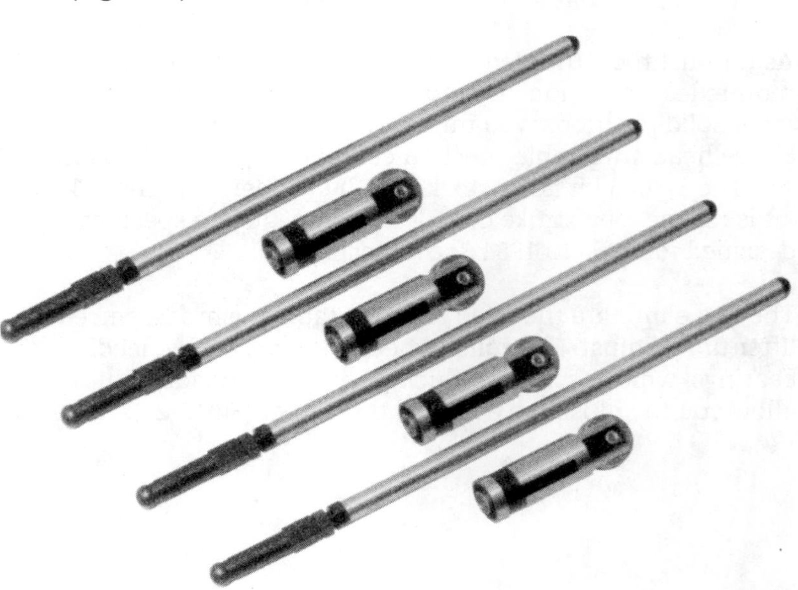

2-29
Crane also produces mechanical lifter sets for older Shovelhead motors.

While on the subject of pushrods, you might consider adding pushrod covers. Sumax makes a nice set to replace the stock uglies. The most elegant pushrod covers are the ones designed and marketed by Arlen Ness. The Ness units are the best made, and best looking, and utilize the factory inner sleeve and spring, a proven sealing method. Arlen's covers do not leak like most aftermarket units. The Ness billet milled-aluminum pushrod covers are sold in sets of four and are available for 1984 and on Evolution Big-Twins, 1966 through 1984 Shovelheads, and 1986 through 1990 Sportsters. All pushrod cover kits employ James gasket seals. Covers are offered in chrome or satin finish (Fig. 2-30).

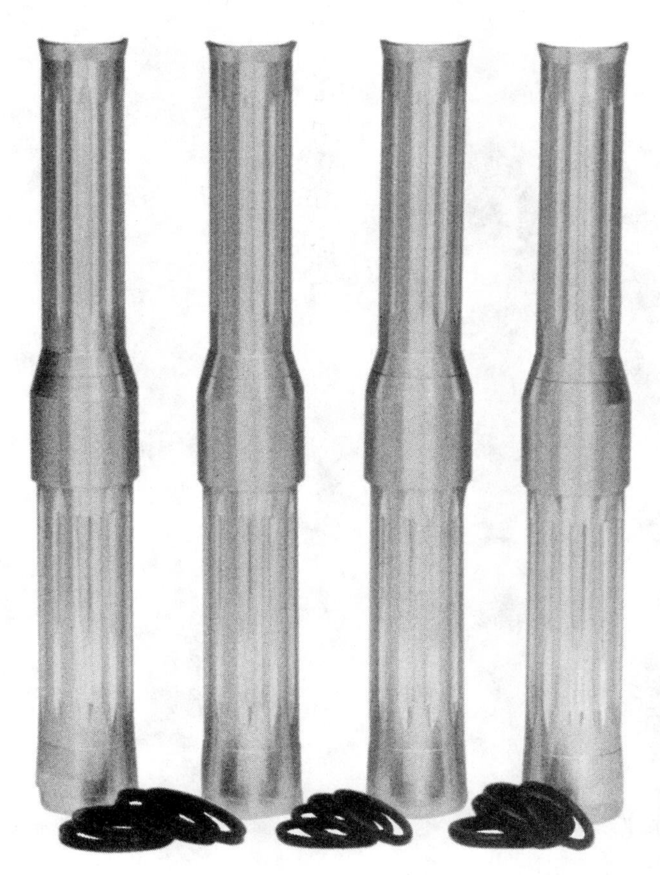

2-30
Arlen Ness billet aluminum Ness-Tech pushrod covers are decoratively machined. The units utilize the factory inner seal for controlling leakage.

CHAPTER 3

Top end modifications

THE GREATEST power gains within the Harley engine, Big-Twin or Sportster, take place in the top end. Basic improvements relative to top end performance can be realized by changing the carburetors and tuning the exhaust. These procedures are detailed in Chapters 5 and 6.

This chapter covers top end modifications to cylinder heads, valves, porting, and compression. The idea is to rework and modify the heads to the point that they will pull 146 to 148 c.f.m. (cubic feet per minute), flow characteristics that make for a highly powerful, potent engine.

Compression ratio

This chapter will cover increasing motor sizes and pushing engines to their limits with bore and stroke increases and high-performance heads. Some procedures may reach the outer limits of the reader's budget, so, you might want to consider a more subtle approach such as merely raising the compression ratio.

What is compression ratio?

Motorcycle engines, and most gasoline powerplants, obtain their power, from the sudden burning of the fuel-air mixture after it has been compressed in the cylinder, just prior to ignition. The mixture flows into the cylinder, the valves close, the piston rises, cramming the mixture into the small volume of the combustion chamber. The amount that the fuel mixture is reduced from its original volume is stated as the compression ratio, 9:1, 10.5:1, and so forth.

In general, this method of compressing the vapor works very well, but in two areas it is potentially troublesome. First, the squeezed mixture has a tendency to explode, rather than burn, so a fuel must be used that inhibits these explosions, called "knock" or detonation. Secondly, the sudden reduction in volume of the fuel-air mixture creates heat, which in extreme cases, can cause ignition before the proper time in the cycle. The problems of detonation and pre-ignition become more serious with higher compression ratios, but with proper engine design and the right fuel they can be minimized.

Measuring compression ratios

The compression ratio of an engine is usually stated in the service literature or the owner's manual, but if not here's how to measure it. First, determine the displacement of an individual cylinder from engine specs, or by measuring the bore and stroke of a disassembled engine. The volume of a cylinder is: (π, or *pi*, × radius^2) × stroke. Once you know the displacement, you can measure how much the mixture is compressed in the combustion chamber. You will need a 100-milliliter graduated *burette*, available from scientific supply houses for about $15, and either some 10 or 20 weight oil or alcohol dyed with food coloring, for visibility.

Advance the cylinder to Top Dead Center by running the piston up as high as it will go, with the spark plug removed. Make sure the cylinder is on its compression stroke, with both valves closed. Tilt the engine so that the plug hole is at the highest point in the combustion chamber. Fill the burette with either the oil or alcohol and note the level by the numbers on the side. Open the tap and see how much fluid it takes to fill the combustion chamber to the bottom of the spark plug hole. By noting how much the liquid drops in the burette, you have determined the volume of the combustion chamber.

The combustion chamber volume, plus the engine displacement, tells you the initial amount of fuel inside the engine before it is compressed. The combustion chamber volume alone is the volume of the mixture after it is compressed. These two volumes are expressed together as the compression ratio. Let's go through this calculation for a Harley 74.

Original cylinder displacement = 37 cubic inches. Convert displacement to cubic centimeters (multiply by 16.5) = 600 cubic centimeters.

Volume of combustion chamber at TDC measured with burette = 80 milliliters = 80 cubic centimeters.

Total cylinder volume (displacement plus combustion chamber) = 600 plus 80 = 680 cubic centimeters.

Ratio of volumes before and after compression = 680:80.

Reduced to lowest terms = 8.5:1.

How to adjust compression ratios

Anything that reduces combustion chamber volume at Top Dead Center will raise compression. The easiest method is to install high-compression pistons, which have higher domes than stock, and reach farther into the combustion chamber. In addition, thinner-than-normal head gaskets can be used to drop the top of the combustion chamber down closer to the piston crown. Sometimes either the lower surface of the cylinder head or the top surface of the barrel can be milled to raise the compression ratio in the same way. A big bore or stroker engine will end up with a raised C.R. because it is taking a larger than stock displacement and squeezing it into the same stock combustion chamber.

Conversely, to lower compression ratios, flat or even dished-top pistons can be used, extra-thick head gaskets, or compression plates under the barrels, can be added, and the combustion chamber or pistons can be machined to increase volume.

Should you change your engine's C.R?

It is difficult to lay down rules about how high a compression ratio can be used with a certain fuel. Especially at high speeds, the amount and density of the fuel mixture that is getting into the cylinder, the "effective cylinder pressure" is as important as the actual compression that takes place. For example, most supercharged engines run low C.R.'s, 8:1 or less, but the actual pressure in the cylinder head is much greater, probably 12 or 13:1, because of the force the blower rams the mixture in with. Also consider the dude with a 74 who lives in Denver or Quito, Ecuador. He might be running 10:1 pistons, but the actual pressure and octane requirement inside his cylinders is likely to be less, because of thin air at high altitude, than an identical engine jammin' across Death Valley. You can see that compression ratio is only a rough indication of how highly pressurized things are inside your engine (Fig. 3-1).

A look at the accompanying charts will give you some idea of where you should be with an average four-stroke motorcycle engine. The figures used are fairly conservative because the high-speed air-cooled nature of bike powerplants makes them especially sensitive to excessively high C.R.s. A watercooled Detroit V-8 will take considerable abuse from detonation, preignition, and overheating; most motorcycle engines won't.

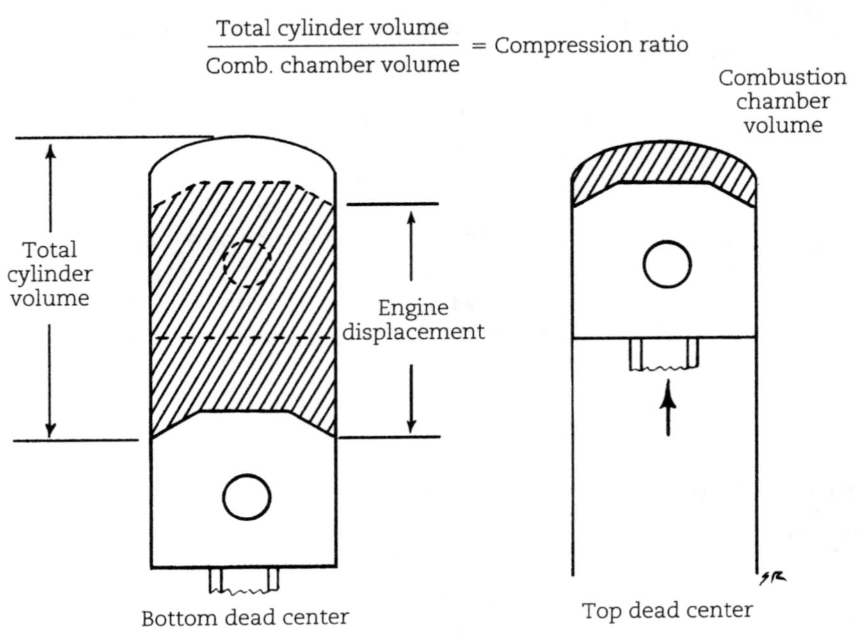

Compression Ratio Chart

Use	Compression ratio
Utility street use (stop and go, extended idling, etc.)	up to 7.5:1
General highway use	up to 9.0:1
High-performance street	up to 10.5:1
Racing	10:1 and over (up to 16:1 on methanol)

Types of Fuel and Compression Ratio Requirements

Fuel	Octane	Compression ratio
Most lead free gas	86-90 octane	up to 8.0:1
Low-lead sub-regular	90-92 octane	up to 8.0:1
Leaded regular	91-95 octane	up to 8.5:1
Middle premiums, "Ethyl"	94-99 octane	up to 9.5:1
Super premiums, lead-free "White gas"	100 plus octane	up to 10.5:1
Aviation gas	120 octane	over 11:1
Methanol blends	120 plus octane	up to 16:1

These charts are only intended as a rough rule of thumb. Because of good combustion chamber design and copious finning, one engine might use a relatively high compression ratio without distress; another might be prone to detonation or overheating with the same fuel and road conditions. Also, octane ratings vary widely, so it is best to determine your engine's own needs by roadtesting, and use a higher octane if in doubt.

3-1 *Compression ratio chart. How C.R. is ascertained and general ratio recommendations.*

The advantages of higher compression ratios are increased horsepower at both high and low speeds and, often, improved gas mileage. The disadvantages are likely to be higher octane requirements for knock-free operation, more heat load on the engine, and harder starting.

Most recent multis come with fairly high ratios out of the crate, however, and much higher ratios may not be a good idea. Because of their copious head and cylinder finning, most can handle the extra heat produced by higher compression better than early designs. The super-premium fuels available eliminate detonation at most ratios. However, the high-domed pistons, necessary to achieve 11:1 or so, may mask valves and restrict breathing to the point that no power increase over stock will result.

How to set up a high-compression engine

If you feel there's room for improvement in your own scooter, are willing to sacrifice a little engine flexibility, and you can stomach the soaring rates for high-octane gas, here's what to do.

Building an engine with just higher than stock compression is relatively easy, but there are a few precautions. First, piston-to-wall clearance must be increased slightly because of the increased horsepower and heat that the engine will be producing. If you are installing high-compression pistons, follow the manufacturer's advice on this. Some high-dome buckets are cast or forged from special alloys that do not expand as much when heated, allowing stock clearances to be used. Because of the likely increase in heat output, it may be a wise idea to open up the top-end rod bushing a thousandth or so, for a looser wrist pin fit. Consider running an oil cooler, too.

Milling the head, or cylinder barrel, to achieve a higher C.R. is not recommended and should only be done as a last resort. If you go this route, take great care that the piston does not actually contact the head, and that the basic combustion chamber design is not changed. For example, squish bands may have to be remachined back to the proper depth if the head has been milled.

Once the high-compression components have been installed, it is a good idea to use some modeling clay to check valve-to-piston clearance inside the engine. In essence, you stuff a big wad of clay inside the engine, turn it over, and measure how close the valves come to the piston top by measuring with micrometers the indentations left in the clay. There should be about .050 inch remaining, if not, *flycut* the pistons to the necessary depth. After flycutting, be sure to remove any sharp edges and, the whole interior of the combustion chamber should be checked for any sharp ridges, points or small burrs. In a stock engine these might not cause problems, but with higher compression ratios they can form hot spots causing preignition, even with the highest octane fuels.

Once all the clearances are correct, assemble the engine and check the actual C.R. with the burette and oil to make sure it is kosher. Increased compression ratios are likely to demand a lot of the engine's induction and ignition systems, so you might want to make a few improvements in these areas as well.

Keep in mind that an excessively high ratio may be worse for your engine in the long run. You might gain some horsepower, but lose engine life when, for example, bearings run hotter and don't last as long. You might gain just as much horsepower by staying with your stock pistons and making some improvements to the ignition, carburetion, or exhaust systems. High-compression ratios require high octane and heavily leaded fuels. The lead that hangs around your engine internals doesn't do any good for spark plugs or valve stems and the lead that goes out the tail pipe doesn't do the world any good either. These days it might be better to drop your C.R. a couple of points, use low-lead gas, and give Mother Nature, and your wallet, a break (Fig. 3-2).

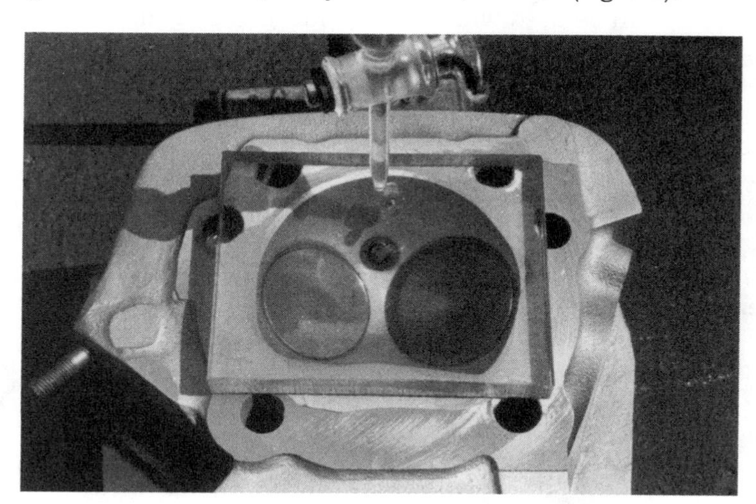

3-2
Combustion chamber head volume can be determined by measuring the heads. To measure the volume, a 100-millimeter glass burette, a flat glass to cover the chamber, and a visible measuring liquid are used.

For substantial changes in compression ratio, special high-compression pistons are the first choice. Since leakage through the rings can create compression loss, you may want to use the new gap-less rings, which can increase operating compression throughout the rpm range. S&S offers a wide range of medium-to-high-compression pistons to accommodate any engine requirement.

High-performance heads

High-performance heads, or reworked stock heads, are a key to high performance. Improving mixture flow to the combustion chamber is a prerequisite for power. The carburetor draws in air and vaporizes the fuel, creating a combustible mixture for igniting and the cam determines when the mixture is fed, via the valves, to the combustion chamber. After ignition, the piston delivers the power pulse, through the connecting rods, to the crank. The head is the connecting link that enables the system to work efficiently. Since stock Harley heads are not created to provide optimum flow, performance buffs may desire the improved aftermarket versions available today.

Rev-Tech (Custom Chrome) offers high-quality aftermarket heads for both Evo Big-Twins and Sportsters. With polished combustion chambers, these heads provide increased fuel efficiency and horsepower. The heads are offered with the capability for using two spark plugs, employing the 12-millimeter plugs found in the Evo Sportster. All stock Harley components can be used with these heads, rocker arms, springs, gaskets, etc. or suitable upgraded aftermarket components. Every Rev-Tech head comes with its own flow spec-sheet. There are minor variances due to hand finishing, but pairs are usually very closely matched.

Installation of these heads on a slightly modified bike carburetor and cam, will provide an instant improvement in performance and fuel efficiency. The heads *flow* about 30 percent more than stock Harley heads at 3200 to 3600 rpm's. Rev-Tech heads are available for all Evolution engines, including 1993 models with the new breather system.

Undoubtedly, the finest aftermarket Evo heads are those manufactured by Lou Trachtenberg's S.T.D. Development Co. of

Chatsworth, California, for both Big-Twins and Sportsters. The S.T.D. V-2 style heads feature improved design characteristics that make them the best flowing heads on the market. The intake port opening is raised ⅜ inch to allow better entry of the mixture to the combustion chamber. With stock sized valves, this change alone increases flow by 20-percent. Porting and polishing, additionally, increases the flow rate. In raising the intake manifold position, material was added so that stock manifolds will bolt on without modification. S.T.D. heads also feature large diameter valve seats for the use of larger valves. A four-bolt, fully machined, exhaust pipe mount is utilized to allow the head to accept stock exhaust systems, but it can also be modified easily to accept larger, flanged-type exhausts. Available also is a .250-inch raised exhaust port exit, which gives substantially more flow than the stock located port.

S.T.D. V-2 heads utilize a bathtub-type combustion chamber, which provides excellent combustion characteristics minus the valve shrouding common with the single squish band-style chamber. Single plug as well as dual plug heads are available. The combustion chamber size approximates stock chamber size to keep compression ratios uniform with the pistons that are being used (Fig. 3-3).

3-3
Carl Morrow's Unlimited Head Package includes porting, polishing, a valve job, and flow testing. Special stainless steel valves are included. Carl's Speed Shop

Carl Morrow uses S.T.D. heads in all his racing applications and features them in his head machining services. Carl's Tru-Flow head service refines the heads for 10-second street bike potential with mild porting and stainless steel valves; installed with springs set up for your particular cam. Seats are cut on Morrow's exacting Serdi machine and the heads are measured for volume and flow-tested by Carl prior to delivery.

The all-out "Morrow Unlimited Head Process" turns S.T.D. cast heads into fine jewels. Each head is ported and polished and the chambers are reshaped. The heads undergo a "Perfect Radius" Serdi valve job topped off with the addition of Carl's special stainless steel valves. The heads are then measured and flow tested to ensure optimum power.

Those of you putting around on 883 Sportsters may want to upgrade to the "wee beastie" level of 1200. Carl's Speed Shop has the answer, offering their CM 1200 conversion kit. Carl Morrow will take your existing cylinders and heads and rework them till you have a street machine capable of running in the low 11s (Fig. 3-4).

3-4
Carl Morrow's CM 1200 Sportster Conversion Kit will add new life and power to an 883 Sportster. The kit features boring and honing, CM cams, S&S pistons and rings, Andrews pushrods, special valve springs, and reshaped intake and exhaust ports. Carl's Speed Shop

The modifications include modified combustion chambers, larger, stainless steel valves, and special valve springs with Titanium collars. Carl's special CM 4 cams are also part of the package. The cams feature a .520-inch lift and a 250-degree duration; the valve travel is set to match the cams. Cylinder work includes boring and honing to accommodate the larger bore S&S pistons.

With this upgrade kit the former 883 will be much quicker out of the hole and much more potent in the mid-to-top-end range. If you are looking for a real performance and power metamorphosis out of your 883 Sporty, Carl's Speed Shop offers the solution.

Shovelhead & Sportster performance buffs can benefit by specialty heads such as the Strocieck heads, which when bolted on, provide noticeable horsepower gains. The fine polished finish of the combustion chamber and specially designed ports assist in providing the most ideal fuel flow possible. The Strocieck milled aluminum *hemi-heads* contain enlarged intake ports and oversize exhaust ports. Spark plug positioning is dead center between the valves, not offset as in stock Shovelheads, providing more even combustion in the head and resulting in a 15-horsepower gain (Fig. 3-5).

3-5
Strocieck heads offer a 15-horsepower gain over stock Harley heads.

Fueling-Rivera four-valve heads

Since the initial step to performance gain is making the engine breathe better, it is obvious that anything that will increase fuel volume intake will enhance performance.

Fueling-Rivera has come up with a working four-valve head and it is rumored that Harley is closely studying the concept. Installing these heads will give a 60-horsepower engine 95-horsepower capabilities. With optional parts also marketed, even greater horsepower gains can be realized (Fig. 3-6). The heads feature a specially designed combustion chamber, with the spark plug hole in the middle of the head. This allows the fuel charge to burn more efficiently. The valve area has been increased 26 percent, which in turn, increases intake flow by 25 percent. This increased fuel flow is produced in part, by a special fairing in the intake port that streamlines the mixture flow to the cylinders. Exhaust ports have also been modified to allow exhaust gases to escape faster.

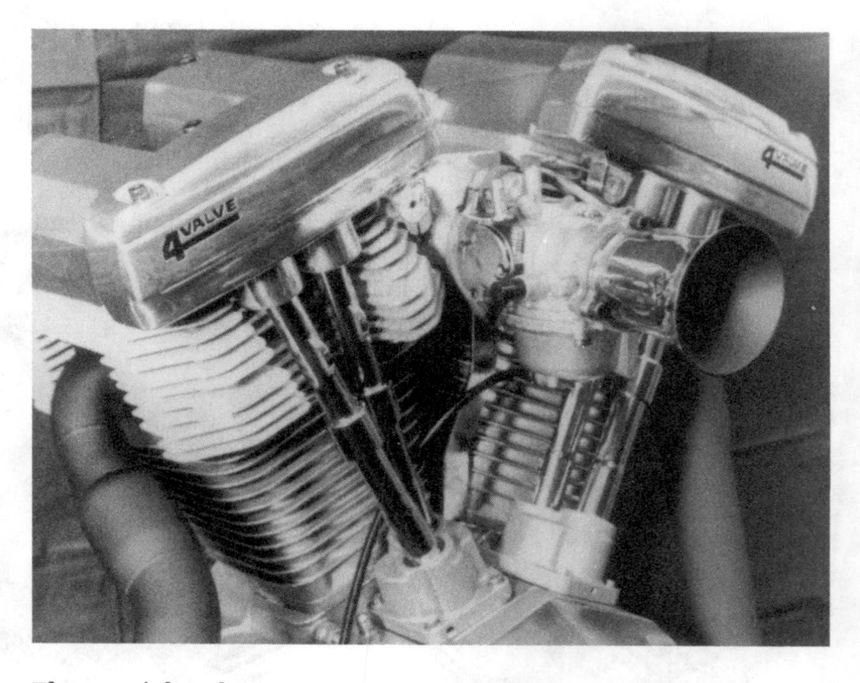

3-6
Fueling-Rivera 4-Valve heads, a step forward in head performance.

The special rocker arms operate on a swivel-foot adjuster, which resembles a ball with a flat spot. When the valves are adjusted, the flat spot remains in constant contact with the valve stem as it travels up and down.

Fueling-Rivera heads cannot be used with stock exhaust systems. Depending on year and model Harley, the applicable exhaust system is available from Rivera. The heads will perform with a stock cam and hydraulic lifters but Rivera recommends the engine not be pushed beyond 5000 rpm unless a set of semi-solid adapters is installed. The heads will work with virtually any carburetion system, and, for optimum ignition, the Dyna-S single-fire system is highly recommended. A special cam is also offered as an option.

Installation is not too complex, but I don't advise that an inexperienced wrench take on the job, even though Rivera does offer an installation video that simplifies the task. It's basically a head change, but there are some intermediate modification procedures that require some expertise (Fig. 3-7).

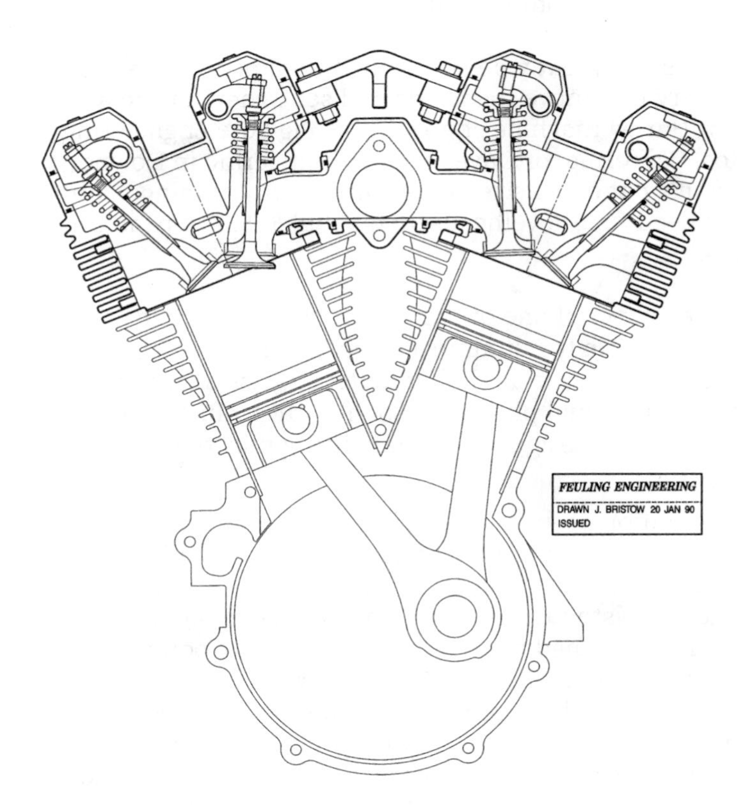

3-7
The 4-Valve system. Rivera-Fueling Engineering

Dual plug heads

Because of spark plug offset, uneven combustion might create pressure pockets in the chamber and cause detonation and possibly "pinging." The partial combustion syndrome is evidenced by the usual one-sided carbon buildup on the tops of the pistons. The benefits provided by dual plugs are a cleaner, smoother running engine, and one that has less tendency to "ping."

Dual plugs are especially effective on older, pre-Evo engines, where it is necessary for the spark to reach into the side of the chamber, opposite the plug. It is not necessary for Evos. With the super turbulence created by the Evo engine, the spark reaches into all the confines of the chamber. The smaller combustion chamber has less area for the flame to travel about, hence, better combustion.

Older Harleys will greatly benefit from a dual plug conversion. It is necessary to remove the heads and send them out to a competent Harley machine shop such as Big Bore Engineering in Cerritos, California, or Dave Mackie Engineering in Downey, California. It's not expensive either, with prices running about $50.00 per head for Shovelheads, and about $75.00 a head for Sportsters.

Jim McClure, owner of one of the world's fastest 96-inch Sportsters, was able to cut almost ½ second off his E.T. with dual plugs. If you go the dual plug route, I advise also going to high performance dual plug coils, such as the units offered by Dynatek and Accel designed for dual plug application.

Big bore cylinders

The augmentation of the stroke in an engine creates longer and more thorough fuel combustion and, consequently, higher torque. It also requires connecting rods of a greater mass and weight and the piston traveling up and down over an increased distance. The combined mass of rods and pistons pounding away within the cylinders can put such a strain on the pinion and sprocket shafts that their holes in the cases can become elongated. Extra long rods also induce piston tilt in the cylinder, which can permit *blow-by* and compression loss. Accelerated piston speed can also promote piston burn, or melting, because the piston does not receive enough oil to cool at higher piston speeds. Many performance specialists believe that a larger bore, with a shorter or stock 4¼-inch stroke, alleviates the aforementioned problems, while still offering a substantial power increase. The power increases in torque with short stroke, big bore engines, however, occur in the higher rpm ranges, which is an acceptable tradeoff.

As stated before, stroking an engine will add more inches but requires longer piston travel. The more gas and air the engine can pump in, the more power it will produce. Big bore cylinders provide more cubic inches, but without the wear associated with increased stroking.

One of the leaders in the big bore field is Axtell. This company is an offshoot of the original Dytch Co., the Dytch barrels of the '60s being the only quality barrels at the time (Fig. 3-8).

3-8
Axtell barrels on Felix Lugo's "Intimidation" an early nitrous fed Sportster.

Axtell's "EV Mountain Motor" cylinders are made of cast iron. They are available for all late Evos and traditional style cylinders are also produced for all the popular Harley engines using base stud mountings. Bore sizes start at 3½ inches and go up to 4 inches. Axtell produces the "Mountain Motor" kit for 883 Evo Sportsters. The "Mountain Motor" combines Axtell's

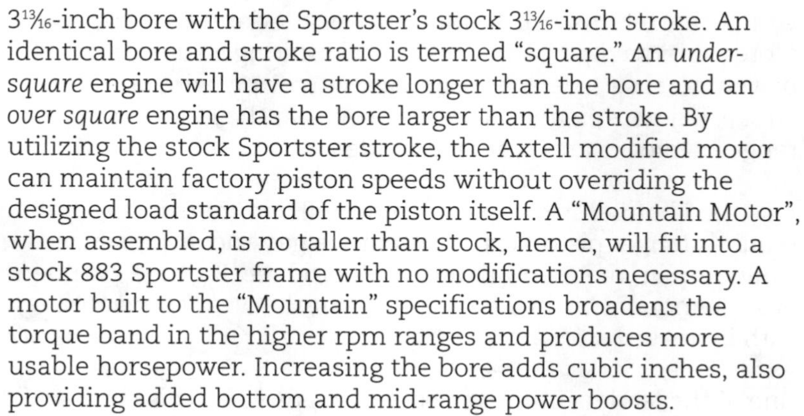

$3\frac{13}{16}$-inch bore with the Sportster's stock $3\frac{13}{16}$-inch stroke. An identical bore and stroke ratio is termed "square." An *under-square* engine will have a stroke longer than the bore and an *over square* engine has the bore larger than the stroke. By utilizing the stock Sportster stroke, the Axtell modified motor can maintain factory piston speeds without overriding the designed load standard of the piston itself. A "Mountain Motor", when assembled, is no taller than stock, hence, will fit into a stock 883 Sportster frame with no modifications necessary. A motor built to the "Mountain" specifications broadens the torque band in the higher rpm ranges and produces more usable horsepower. Increasing the bore adds cubic inches, also providing added bottom and mid-range power boosts.

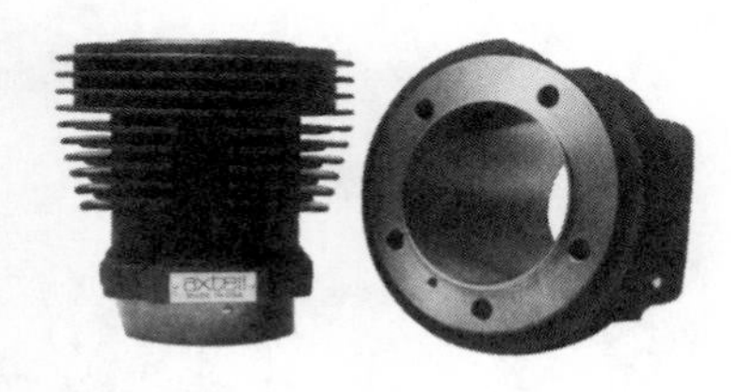

3-9 *Barrels for the Sportster and Shovelhead are available from Axtell, with the Traditional Axtell look, for stock up to 4-inch big bore, strong and bulletproof of 80,000, p.s.i. ductile iron.*

With virtually all aftermarket overbore cylinders that exceed stock specs, the engine cases must be re-radiused to accept the wider cylinders. Special 16830-72 short studs replace the stock cylinder studs. For more information on the Axtell "Mountain Motor" kits, contact the company or obtain one of their catalogs (Figs. 3-9, 3-10, 3-11).

3-10 *Axtell "Mountain Motor" cylinders for Evos feature internal oil drains in stock, $3\frac{5}{8}$-inch, and $3\frac{13}{16}$-inch big bores cast iron for extra strength.*

Hyperformance has been around since 1983, providing machining services including valve guide installation, cylinder head reconditioning, combustion chamber remachining, and professional porting. They also offer their Hytech line of quality products.

Hyperformance "Big Jugs" are cast from #80-60-03 *ductile* iron, heat treated to 80,000 p.s.i. for maximum strength, and then precision machined to uniform wall thickness. Street cylinders feature machined cooling fins with oil passages machined and threaded to accept aircraft-quality external oil lines.

3-11 *Finless cylinders are also produced by Axtell for drag and race applications for Evos and Shovelheads.*

Hyperformance will also provide any case/cylinder head combination and any bore size up to 3¹³⁄₁₆ inches for street machines (Fig. 3-12).

3-12
Ductile iron "Big Jugs" are available in kit form with machined fins for the street. Kits are complete with custom machined Hytech pistons, rings, teflon buttons, Viton "O"-ring head gaskets, and high tensile strength studs and bolts.

S&S, a long-time leader in aftermarket performance goodies, offers a wide variety of big bore as well as stock cylinders. For Evolution V-2 Big-Twins they offer their 3⅝-inch bore Sidewinder kits, which feature the big bore barrels, 1700 or 1900 series pistons with rings, wristpins and retainers, and head and base gaskets. Stroker flywheels, connecting rods and shafts are optional. Additional crankcase boring is required for proper fit of the big bore cylinders.

For Evo Sportsters, Sidewinder kits are available in bore sizes of 3½ inches and 3⅝ inches, designed to fit all 883 or 1100 engines. Cylinders and pistons in the big bore package are all that are needed to make a stock stroke, 3⅝-inch bore, 79-cubic-inch V-2. Modifications required to install an S&S Sportster V-2 Sidewinder kit are similar to those required to install Big-Twin Sidewinder big bore kits. All 883 installations must have cylinder head modifications, which can be done with a hand grinder.

Trock Cycle produces ductile iron big bore cylinders for any bore and stroke combination without the need of stroker plates. They also market 3⅝-inch big bore and Shovelhead cylinders. Their big bore $3\frac{13}{16}$-inch cylinders, for Shovelheads, Panheads, and Evos, utilize the standard base bolt mounting system.

Sputhe Engineering has helped make big bore history with their line of "Nitralloy" big bore kits (Fig. 3-13). "Nitralloy" barrels are cast from 383 aluminum alloy, possessing a *tensile* strength of 45,000 p.s.i. It is cast over a lascomite sleeve and injected into a steel die at over 5,000 p.s.i., ensuring a perfect bonding to the sleeve. The cast-in liner is 60 percent thicker than stock for increased rigidity. The "Nitralloy" barrels have greater fin area and a symmetrical fin pattern to reduce thermal distortion. Crankcase cylinder holes must be opened up in order to accommodate the wider cylinders.

3-13
Sputhe Engineering's Nitralloy big bore kits for Big-Twin Evos that increase the engine's displacement to 95.4 cubic inches.

Sputhe offers a 95.4-cubic-inch Big-Twin kit and a 104-cubic-inch Big-Twin kit. The 104-inch kit uses the same length cylinders as their 95-inch kit, the difference being the pin location in the piston. The 104-inch kit must be used with 4⅝-inch stroke flywheels. S&S 2060 flywheels are recommended for this purpose.

Big Bore buildups are not necessarily complex, provided there is no machine work to be done to the cases. In most instances, a big bore swap can take place while the engine is still in the frame. We will show how a big bore installation is undertaken. The cases are Delkron and the replacement cylinders Sputhe 3.780-inch big bores (Fig. 3-14).

3-14
Even though big bore cases are available, some of them will not accommodate the excessive bores offered by Sputhe, Axtell and others. Some machining is required. Here John Sachs bores out Delkron cases to receive the Sputhe 96-millimeter bore cylinders.

After the cases have been drilled out, if required, assemble the lower end and attach the pistons. If you are not opening up the cylinder holes, leave the engine in the frame; remove only the heads, stock pistons and cylinders. The steps are simple and described in the Harley shop manual or the Clymer mechanic's manuals. With the head and cylinder removed, rotate the engine so the piston rises up out of the crankcase to its base level, making the newly mounted big bore piston accessible (Fig. 3-15). Next, install the rings onto the new piston in the positions and order as instructed by the piston or big bore kit manufacturer (Fig. 3-16).

3-15
Pull the piston out of the case by rotating the motor until it is out far enough to be inserted into the barrel.

3-16
Install the rings in their respective grooves, with gaps staggered as recommended.

Now it's time to mount the cylinder barrel. Make sure you use a new and correct base gasket. Big bores tend to blow gaskets more easily than stock motors. Paper or composite gaskets are not totally reliable. I find that the best gaskets available for big bores are the ones manufactured by P.C.I.M., Inc. These strong, but malleable, copper gaskets just don't blow out and are available in various thicknesses. P.C.I.M. base gaskets, as well as head gaskets, are manufactured for all types of big bore cylinders. These are the most reliable gaskets for this type application (Fig. 3-17).

3-17
P.C.I.M. copper gaskets are the best for high compression engines, not prone to disintegrate under pressure.

Prior to slipping on the barrel, compress the rings so that the piston will slip easily up into the barrel. The best way, to protect both rings and cylinders from damage, is to use a ring compressor (Fig. 3-18). Remove the ring compressor after the rings are fully in the cylinder sleeve. Slide the barrel over the dowel bolts until the base is firmly seated on the base gasket, which should be positioned and sealed in place (Fig. 3-19). In Fig. 3-20 the piston is in place after the cylinder has been seated on the base gasket. Check to be sure that the piston has unrestricted travel prior to installing the heads (Fig. 3-21). Do this by rotating the crankshaft with a wrench. Install the head and torque it down to the required specifications (Fig. 3-22).

3-18
Use a ring compressor to contain the rings so that the cylinder can be slipped over easily.

3-19
The barrel is slipped over the dowel retainers and over the piston.

3-20
If the rings are compressed properly, the piston will slip into and up the cylinder.

3-21
The barrel snuggled into place, ready for the head to be put on.

3-22
The head is torqued down. All that is left now are the rockers and rocker covers and the job is complete.

Do not under- or over-torque cylinder head nuts. If under-torqued, the crankcases or head may leak and if over-torqued the cylinders may be distorted. Installed properly and maintained, big bore kits will provide you with trouble-free, high power operation. In Fig. 3-23 you can see the difference

3-23
At top is the stock Harley piston dwarfed by the Sputhe Piston in the Sputhe barrel.

between the stock Harley cylinder and the big bore replacement. Barrel and piston are from Sputhe.

Performance valves

One primary engine hop-up is the providing of greater intake fuel volume and density by means of oversize valves and by slightly enlarging and reworking the intake passages.

When oversize intake valves are installed, exhaust valves are usually modified, since larger exhaust valves allow the engine to expel burned gases in greater volume. Though the addition of oversize valves does much to improve Harley breathing, simply installing bigger valves may offer little or no performance gain. When installing oversize valves, it is recommended that the intake ports be enlarged in diameter, to correspond with the increased valve diameter. The objective of bigger valves is to permit enlarging the width of the intake ports, allowing greater amounts of vaporized fuel to enter the combustion chamber. The passage enlarging procedure is commonly referred to as *porting*. Obtaining an increase in mixture can easily be achieved by enlarging the ports, but keep in mind that one can go overboard in the process. A too-large port may work in reverse, actually slowing down and inhibiting mixture velocity. The area in which particular care should be exercised is between the valve guide and seat.

Before undertaking oversize valves, be sure to obtain proper porting specifications from a Harley dealer or have the dealer perform the porting job. The ports should also be polished, since a better, unrestricted airflow will result, with less turbulence due to rough spots or surface irregularities. Since the valve seat aids in directing airflow, it is important that the seat surface is smooth and free of sharp edges or obstructions that cause airflow to be impaired (Fig. 3-24).

3-24 *Warr stainless steel valves by F.C.C. feature a special contour in the stem to promote better airflow. These valves are available in various sizes, stock and over.*

The most common oversize valves for Sportsters have been the ones manufactured by Harley-Davidson. These are outdated, and I have discovered what I consider to be the optimum replacement valves. These oversized, improved valves are currently manufactured by Manley Performance Products, Bloomfield, New Jersey, long-time leader in the automotive performance goodie line. The Manley valves are superior to the

Harley parts in many respects. They are of one-piece construction, as opposed to the old Harley two-piece construction. Also, they are made of stainless steel and chrome-plated, a feature that aids in preventing galling. The Manley valves allow improved airflow, since they are specially contoured and sized. Also, lighter in weight than Harley valves, they offer greater rpm potential. Installation of the newer, improved Manley valves will afford a power gain of 3 horsepower or more.

The machining procedure involved in installing oversize valves is relatively simple, if one has the proper tools. Necessary tools include: 30-, 45-, and 60-degree seat cutters, a *pilot* to align and center the cutters, a fly cutter, and some good lapping compound. The machining procedure is illustrated.

If you're an above-average wrench, you might want to take on the job yourself. Either grinding stones or steel cutters may be used to do the job.

Carbide stones are the older method and, though they serve to do the job, they are not as accurate and do not cut as fine a seat (Fig. 3-25). As for technique, experience is the best teacher. If

3-25
Stone grinders are an acceptable method, though not as popular as in the past.

you want to do it yourself, you should first practice on an older junk head before turning loose on a mint Evo head. Make sure you have the proper tools to produce the right radii and angles.

Neway has an excellent valve seat cutting kit. They even market special Harley kits to cut the necessary three angle seats (Figs. 3-26 and 3-27). These cutters are excellent on alloy heads. Alloy heads do not fare too well under carbide stones, which tend to load up and clog with material under heavy use.

3-26 *Neway valve seat cutting kits feature cutters for all necessary angle cuts.*

3-27 *The Neway cutters are hand powered, guided by a pilot shaft equal in size to the valve stem, which is inserted into the valve guide. The kit features varied angle cutters and various width pilot shafts.*

The cutting procedure is relatively simple. You must keep in mind that you will be using 30-, 45-, and 60-degree cutters for cutting the new seat. The 30-degree cutter for the outer relief cut, the 45 for the actual contact seat, and the 60-degree for the inside relief cut. Insert the proper size pilot into the valve guide. Place the 45-degree cutter over the guide and secure it to the cutter hex hub in the T-handle, supplied in the Neway kit. To cut the seat, turn the cutter clockwise, applying enough hand pressure to cut the metal. Always maintain an even downward pressure over the centerline of the pilot. Follow the same procedure in cutting the bottom, narrowing and top, narrowing cuts. (Figs. 3-28 and 3-29).

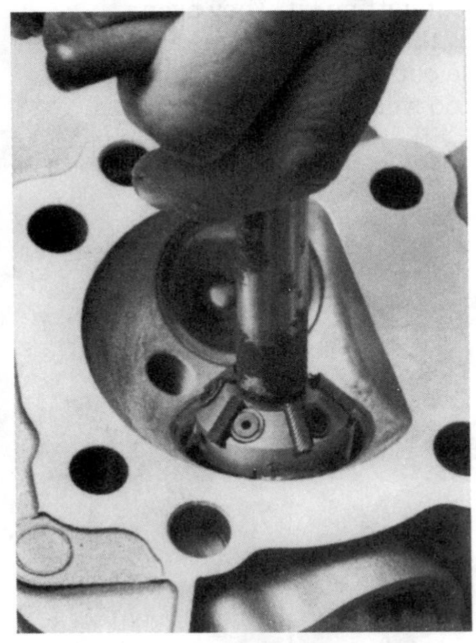

3-28 *Seats are hand-pressure cut using 45-, 30-, and 60-degree cutters.*

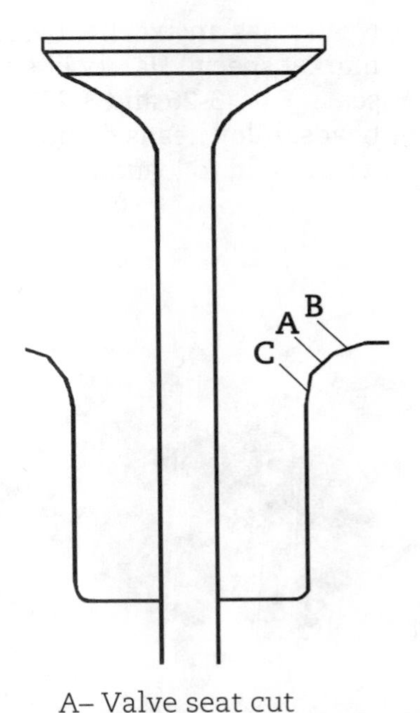

3-29 *A, B and C indicate the various angle cuts on a valve seat.*

To finish, place the valve in the head and check for fit and sizing. If all is copacetic, brush some valve lapping compound onto the seat and *lap* in the valve to fine finish the valve seat (Figs. 3-30, 3-31, and 3-32).

If you are looking for the optimum valve job . . . "Go west!" West to Carl's Speed Shop where the "guru of go," Carl Morrow, does the best Harley engine work in the country. Carl can provide high-tech valve jobs by means of his Serdi 100 self-centering valve seat machine. Carl uses the Serdi machine to get higher efficiency from the Harley engine and has himself developed

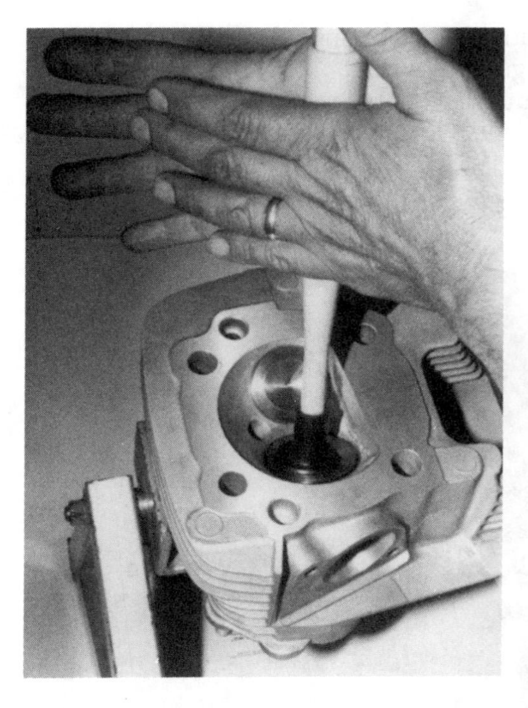

3-30
The valve seat is lapped using a lapping tool, lapping compound, and a back-and-forth rotary motion.

3-31 *Here is a perfect, even, valve seat, the refined machine work of "Monster Motor" man John Sachs of Sunrise, Florida.*

3-32 *The completed valve job with enlarged valves in place and perfectly seated.*

several valve seat cutters that optimize flow through the Evo ports. One of the major problems with the older, three-cut grinding method was proper alignment. The pilot used in the conventional method flexes in the top part of the valve guide making the procedure unreliable and inconsistent. The Serdi machine guarantees alignment with a precision spindle in an air-floating sphere layout. This enables seat cutting with critical reference to the valve guide (Fig. 3-33).

3-33
Carl Morrow, the master himself, at work on his Serdi 100 valve seat cutting machine. Carl's Speed Shop

Porting & polishing for power

There is no question as to the benefits derived from porting or enlarging the cylinder head ports. However, the case for polishing is not quite as clear-cut. One group claims that infinite polishing promotes such free, unobstructed flow that the fuel does not properly atomize. The other feels that rougher walls, in the manifold and ports create more turbulence, which promotes better atomization. The majority, though, admit that polishing does assist in improving mixture intake in to the engine.

To perform porting and polishing we must rely on hand-motor tools such as the Foredom, Dremel, or Black & Decker. For the greatest flexibility in machining and polishing, the Foredom has to be the leader of the pack (Fig. 3-34).

Cylinder heads are mass-produced using casting techniques that leave much to be desired when it comes to fine finishing, particularly on the inner surfaces of the ports. Port surfaces are

3-34 *The Foredom flexible shaft motor tool is ideally suited to porting and polishing. The hand piece, shown here, is compact and will easily access Harley heads and ports.*

usually irregular and pitted. This pitting shows up easily under intense magnification. Pitting looks like a series of craters, pockmarks, ridges, and so on. These irregularities promote turbulence within the ports that inhibit and disturb the free flow of vaporized air-fuel mixture.

By polishing, you allow the mixture to flow smoothly and unrestricted. This is where the controversy enters. Some say rough walls help break up the gas into vapor so that it more easily mixes with the air in the ports. This might be true to a certain extent. But for high-performance operation, there seems to be no question that smoothing out the flow does help the power output.

There's no question that porting is beneficial. Porting increases the diameter of the intake port, increasing the amount of mixture crammed into the engine at any given time. Also, of course, the greater the volume mixture in the combustion chamber, the more power the engine produces. It's really a question of volumetric efficiency.

A rheostat-type control, to vary the speed of the hand-motor tool, is handy and makes the tool more controllable for delicate grinding and cutting. The grinding bits, or "lollipops," used in the tools for fine finishing are also important. Buy the best kind, regardless of expense, as cheap counterparts do not hold up well and might even fly apart under heavy usage. Carborundum grinding bits are available in all sizes and shapes and some are specially designed for porting and polishing.

To open or polish a port, start with a 1-inch "lollipop" in your grinder or motor tool. Insert the revolving grinder into the port opening and begin to smooth out the walls using a slow, circular, clockwise, motion around the port. When you really get deep into the ports, be extremely careful that you don't overgrind around the valve guides. Also, be very careful when polishing or grinding around the valve seats. For small areas or confined corner work, you can obtain small bits that go down to ¼ inch in size. To know how much material you can grind away around the valve seats and guides, consult the Harley manual or check with the local Harley head shop (Figs. 3-35, 3-36, and 3-37).

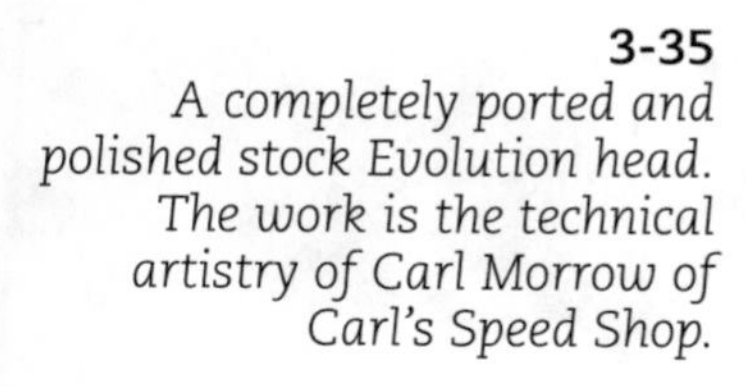

3-35
A completely ported and polished stock Evolution head. The work is the technical artistry of Carl Morrow of Carl's Speed Shop.

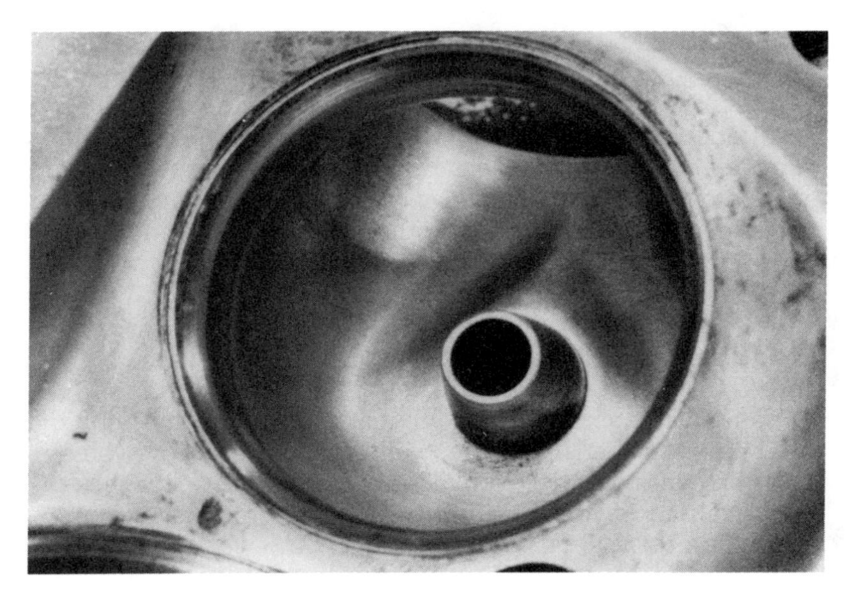

3-36
Closeup view of the port shows the fine contouring of the port areas adjacent to the valve guide.

3-37
Another view of the port contouring.

Valve springs

Heavy-duty, or aftermarket valve springs, deserve more attention, since they are an important part of the valve train. Stock valve springs are a compromise at best and when high-performance head work is undertaken, performance valve springs should be included. Even today's stock, high-output Harley engines require more attention to valve springs. Spacers can be inserted to increase spring tension but the real solution is stiffer springs, particularly with higher lift cams and related components that stress the valve train. Valve springs without adequate tension promote valve bounce, resulting in damaged valves or seats or both. On the other hand, valve springs with excessive tension can overload the valve train and result in added wear on pivot and contact points.

In addition, valve-spring height must be measured and corrected. This also helps in maintaining correct spring pressure. Valve spring testers are marketed to check height and pressure characteristics so you will know if the springs you are installing are up to your applications standards. Check the specifications of your cam and valve train components from manufacturer's specs.

Crane offers their 5-1003 Thermo-Cool springs for 1948 to 1984, Big-Twins, Evo Big-Twins, and XLHs. The Thermo-Cool springs eliminate hot spots in the coils with a special coating that disperses heat and reduces friction. Crane offers two styles of retainers for high-performance needs. Chrome moly 4140 steel is used to provide maximum strength and durability in their steel retainers. Crane's optional, titanium retainers combine strength with light weight to meet performance needs (Fig. 3-38).

3-38
Crane offers valve spring kits with either steel or titanium retainers.

Leinweber offers heavy-duty, triple spring kits for all Evolution Big-Twins, featuring titanium top collars. The racer's edge, these springs, with precision machined bottom collars, are also heat treated and *black oxided*. Leinweber also offers a line of Shovelhead double valve spring kits that include steel split-keepers and shims (Fig. 3-39).

3-39
Leinweber triple valve spring kits contain titanium upper collars and steel bottom collars for evolution Big-Twins.

Pistons

The piston is the part within the internal combustion engine most affected by the combustion process. The piston overcomes inertia and is pummeled into motion by the ignition of the mixture as it is drawn into the head. Connected to the piston is the rod, which, in turn, is connected to the crankshaft serving to convert the linear motion of the piston into the circular motion of the flywheel.

The piston receives the largest portion of heat and stress in the internal combustion engine. The ignition explosion thrusts the pistons downward via the connecting rods, turning the flywheels. The flywheel motion is transferred via clutch and chain, or belt, drive to the rear wheel to provide locomotion. The piston does its job within the cylinder, between the head and crankcase. The combustion energy that acts on the piston would escape around the piston if there were no piston rings. The rings are held in place in grooves in the upper part of the piston. Two of the rings serve as compression rings and the third as an oil ring, to wipe oil on and off the cylinder walls as the engine is running. The compression rings maintain a seal to minimize compression chamber blow by. Cylinder wall-to-piston clearances can be as small as .003 inch or can go up to .015 inch or more on higher compression engines. How pistons are manufactured and to what tolerances is determined by material, design, compression ratio, and intended fuel. High-performance racing engines run much hotter than stock street engines and require pistons designed to survive the added stress.

Through the middle of the piston, along the centerline from left to right, is the wristpin hole, which houses the wristpin. The wristpin connects the piston to the rod. Reliefs machined in the pin allow the installation of wristpin keepers which serve to secure the pin to the piston. In the piston's oil ring groove are small holes that permit oil from the crankcase to rise to the oil ring while the engine is running. In this manner the cylinder wall receives oil with every piston stroke. This promotes both cooling and lubrication.

Stock Evolution pistons are forged, as opposed to cast. Machining of holes, ring lands, and reliefs is done after the forging. On the piston *crown* or top, are two half-moon indentations, machined into the surface for valve clearance throughout the rpm range. The flat-top Evo pistons differ from Shovelhead, Panhead, and Knucklehead pistons, since the Evo combustion chamber is the squish type. The older-type pistons have a domed top, more appropriate to a hemispherical combustion chamber.

Bored and stroked engines rely on special pistons guaranteed to perform at optimum potential and there is a multitude of high performance forged pistons on the market. Sources for these pistons include Axtell, Rivera Engineering, Drag Specialties, and Custom Chrome.

Axtell's iron head and Shovelhead pistons, for example, are forged aluminum with various dome heights to achieve almost any compression ratio. Evolution pistons are also available (Figs. 3-40, 3-41, and 3-42).

3-40
At center is the Stock Evo replacement piston by Axtell, at left, the Axtell high-dome, high-compression piston and at right, the Axtell Pro-Street angle top piston, for the rider who desires the stock look but added performance.

Head
Piston
Cylinder
Squish
.030 – .050
between head
and piston

3-41
Axtell Pro-Street angle pistons of stock bore and stroke, produce a 10:1 compression ratio with a 30-degree angle cut into the cylinder head. Turbulence is high, but the in and out paths are better with a 30-degree rather than the stock 90-degree shelf, and the mixture is forced directly towards the flame front. Horsepower gains up to 30 horsepower can be realized. These pistons are available in 3½-inch stroker version and 3⅝-inch big bore version.

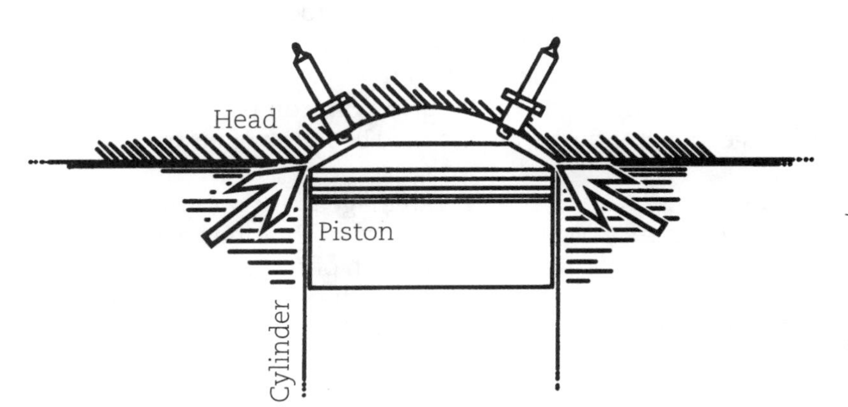

3-42
The Axtell PE535-42-DS high-dome piston has a full 30-degree dome and is designed for a twin shelf head or bathtub-style chamber, such as the S.T.D. or Branch #4 head. Compression ratio is approximately 10:1. Available only in stock bore and stroke.

Roller Rockers

Proper rocker arm geometry is essential for fault-free improved valve train operation (Fig. 3-43). Results of studies by such companies as Rivera Engineering and Crane Cams show that obtaining the optimum in rocker arm alignment greatly increases power potential within the Harley engine. Due to the somewhat archaic design of the V-Twin, obtaining perfect rocker geometry in a stock motor would necessitate four distinct rocker arm designs incorporated into one. When a cam with higher lift is introduced, problems relative to the relationship between the cam, rocker arms, and valve springs can become pronounced.

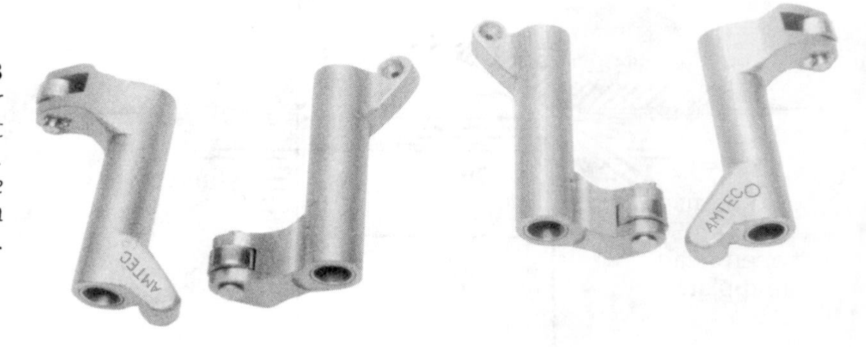

3-43
Rivera Engineering roller rockers are investment cast and hardened to 60 to 62 RC. Stock bronze bushings are utilized so the customer can use the existing shafts.

When a higher lift cam is employed, increased motor output is expected. When operating under ideal rocker arm geometry conditions, cams perform best. With nonconforming rocker arm geometry, full potential of a high-performance cam cannot be realized. Unless the rocker arms can compensate for the change in cam lift, the results may be a loss of power.

Changing the ratios can cause damage to rocker arms and wear to valve guides as well as the distortion that may occur to valve springs due to friction. The reduction of friction by applying proper rocker arm geometry can produce extra horsepower. Aftermarket roller rockers greatly reduce side loads on valve stems and guides.

Roller-type rockers feature roller bearings for smoother operation in addition to roller contact tips (Figs. 3-44 and 3-45).

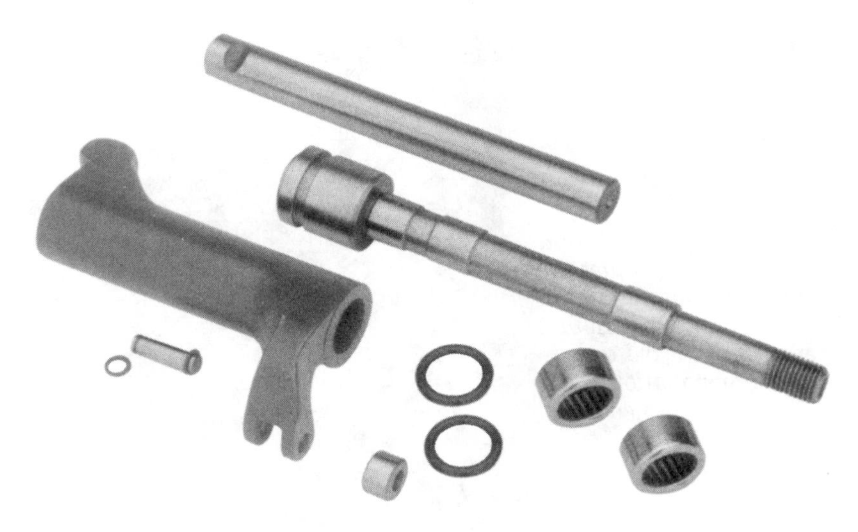

3-44
Crane's excellent roller rocker kit of alloy steel features needle bearing supported shafts. The roller tip reduces valve stem tip wear and prevents galling.

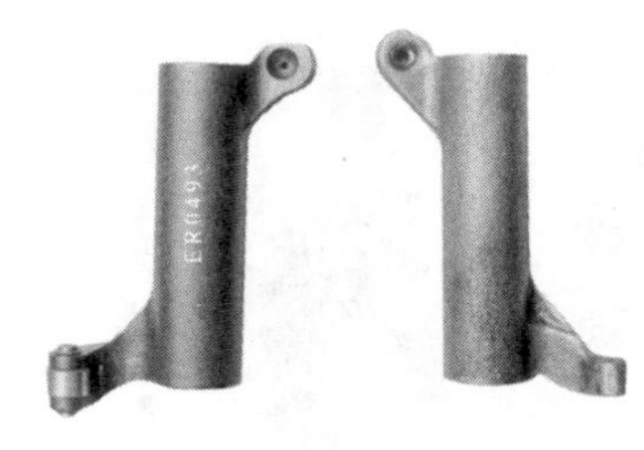

3-45 *Right, a stock Harley rocker arm, and left, the Crane roller rocker with roller tip.*

Roller rockers are fairly easy to install and the procedure can be performed with the engine in the bike. Remove the tanks to allow access to the tops of the heads. Prior to removing the rocker box covers, rotate the engine by the rear wheel with the bike in gear a few times to bleed off the hydraulic lifters (Fig. 3-46).

Remove the rocker cover assemblies (Fig. 3-47). The new roller rockers must be installed one cylinder at a time, with the cylinder being modified at T.D.C on the compression stroke.

Remove the two 5/16-inch bolts that hold the rocker arm shafts, one bolt per shaft, and slide the stock shafts out from their stands. On hydraulic lifter equipped engines, it might be necessary to loosen, but not remove, all the bolts under the

3-46
Loosen and adjust the pushrods for maximum valve lash to facilitate removal of the old rockers and installation of the new ones.

3-47
Remove the rocker box covers to expose the rockers.

rocker box cover at least three turns to relieve pressure on the rocker arms and shafts. This will allow the rocker box cover base to move slightly. Once it is free, you can easily remove the stock shafts and arms. It is a good idea to replace the lower cover gasket if cover bolts are loosened (Fig. 3-48).

3-48
Loosen the rocker box, slide the shafts out, and remove the stock rockers.

To prepare the new Crane rocker arms for installation, remove the rocker shaft and lubricate the bearings, using the Crane assembly lube provided. The shafts should be removed one at a time and the rocker and shaft kept matched. Lubricate and reinstall the O-rings in the ends of the rocker arms. Install the new rocker arm into the box, sighting through the shaft mounting holes and bearings to align the arm. Make sure the pushrod is in the rocker arm's pushrod seat (Fig. 3-49).

Position the rocker shaft so that the notch in the shaft is at the pushrod end of the rocker arm, properly positioned so the locking bolt can be installed (Fig. 3-50). Perform the same procedure for the other cylinder head. If it was necessary to loosen all the bolts under the rocker arm cover, the cover base must now be realigned. Replace the lower rocker cover gasket, if necessary, following the procedures in the Harley service manual. Position the cover so as to center the roller tips of the rocker arms on the valve tips and re-torque the bolts to their proper specs. Install the lock bolt in the rocker arm shaft notch, torquing it to 15 to 18 foot/pounds. Check to see that the pushrod is properly seated. With a mechanical lifter equipped engine, readjust the valve lash. Reinstall the rocker box covers and torque the bolts to 10 to 13 foot/pounds, for 5⁄16-inch bolts.

3-49
Install the rockers in the box, insert the rocker shaft, and align the assembly.

3-50
Position the rocker arm shaft so the notch in the shaft (indicated by finger) is at the pushrod end of the rocker arm.

With cam lifts of over .500 inch install roller rockers and also low friction valve guides, which allow the valve train to act as if it is higher than it actually is.

Rivera adjustable rocker arms

According to Rivera Engineering, all the problems due to incorrect rocker arm geometry have been eliminated with their new adjustable rocker arm and rocker geometry assessment and correction kit. The old method was to add lash caps, but there is not always room for them, especially if stock valves, retainers, and keepers are used. The most effective way to arrive at correct rocker arm geometry is with the Rivera kit, featuring adjustable rocker arms with swivel feet and a special adjustment gauge to set proper geometry (Figs. 3-51 and 3-52).

Rivera has conceived and markets a new pushrod gauge tool for use with their adjustable rocker arm kit. This new item provides the best method for adjusting pushrod length (Fig. 3-53). When the pushrod length is determined, proper adjustments to valve stems are easier or simple adjustments to the Rivera adjustable rocker arms can be made. To use the Rivera pushrod gauge, available optionally, the Evolution rocker boxes are removed and the special bracket and gauge are installed (Fig. 3-54).

To gauge the rear head's intake valve, the scale is shifted to the rear intake position. To gauge the exhaust valve, the scale is

3-51
The new Rivera adjustable rocker arms contain adjustable swivel feet for right on geometry adjustments.

3-52
Cutaway view of rocker box and adjustable rocker arms displays the elements of this new rocker arm design. The arms are hollow to feed oil to the valve springs and valve stem.

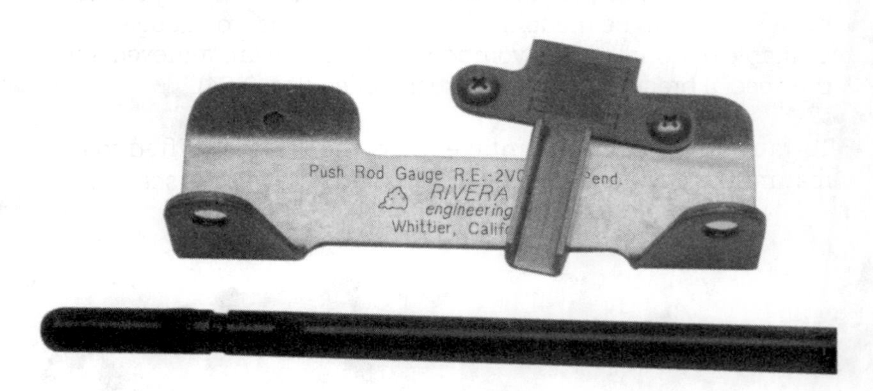

3-53
The Rivera pushrod gauge, a great step forward toward more correct valve train adjustment.

then shifted to that position. On the scale, the zero lines indicate overall length of the pushrod at half-lift and squared position of the rocker arm. Revolving the camshaft will allow the gauge to move up and down in a guide adjacent to the metered scale, marked in .025-inch increments. This allows you to determine the exact lift of the cam. A gauging rod requires adjustment, so that it moves an equal distance above and below the zero line. When the exact location is set, the gauge is

3-54
The special bracket and gauge installed in an Evo rocker box to determine the geometrically correct pushrod length.

put aside and its measurement transferred to the pushrod being used. The functioning length of all four pushrods differ, so all pushrods should be gauged separately to achieve the proper balance.

Modern accessory ignition systems

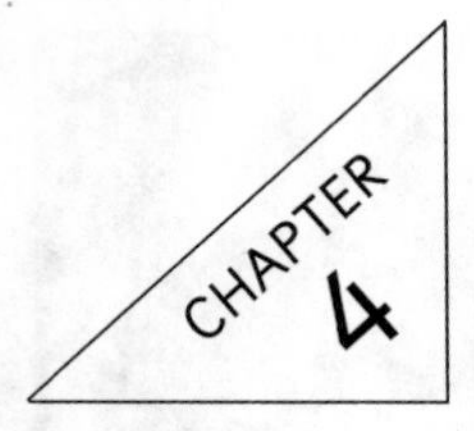

TO INCREASE the efficiency of high-performance modified Harley engines, you need a sure-fire ignition system. Today's state-of-the-art accessory ignitions guarantee high efficiency and dependability.

Starting is the first element in ignition performance and here is a tip picked up from an *Easyriders* article by Jim Thompson about cranking circuit modification. You might find yourself out in the boonies trying to crank over your big stroker, only to be rewarded by electro-mechanical grunting and groaning.

An inexpensive and effective way to overcome this problem is by rewiring the cranking circuit by means of a simple wiring change. By rewiring the circuit in this manner, you can turn the engine over with the ignition off until the engine attains a good cranking speed. Then the switch is flipped to ON and the engine will fire up easily. The engine runs slightly faster with the ignition off and, at low cranking speeds, the flywheels will gather more momentum, making starting less sluggish.

Figures 4-1A through 4-1D show the wiring modification to 1973 through 1981 XLs, FXs, and 1972 through 1981 FLs, done at the handlebar switch housing.

4-1
A) the handlebar switch; B) the factory wiring; C) cutting the wire which is then trimmed and resoldered; D) the new terminal position.

Switches on 1982 and later models may differ slightly, but the modification is similar, since Harley utilizes the same wire color-coding on through the more current models. Figure 4-1B shows the OEM wiring circuit. As in Fig. 4-1C, the wire going to the RUN terminal is cut. After cutting, strip the end and resolder it into the switch terminal, which has a grey wire already attached (Fig. 4-1D). The starter switch is now LIVE whenever the ignition switch is turned on.

This simple modification will provide a noticeable increase in cranking speed. Installation of a heavy-duty starter will also greatly aid engine turnover, but is more expensive.

As a word of caution, by performing this modification you will have disabled the safety feature of the switch. The starter will always be on and will engage *any time the ignition switch is turned on*. Pushing the starter button at any time will engage the starter and crank over the engine. Keep the transmission out of gear when parked as a safety precaution.

Harley engine performance can be greatly enhanced by adding aftermarket components to the stock ignition system, even though the OEM systems are very reliable. A relatively inexpensive addition is the selection of a high-performance electrical system. Key factors in performance are the quality and intensity of the spark discharged by the spark plug. Spark intensity is dependent upon the amount of voltage and the duration of the spark. Also critical is the timing of the spark. Higher voltages and improved firing characteristics of aftermarket ignitions enable the mixture to ignite faster and burn more evenly. The voltage requirement for the normal operation of the Harley engine is about 20,000 volts—the level required to produce a proper spark at the spark plug's electrode. Higher spark voltages will intensify combustion, providing horsepower gains. Conventional Harleys run on 12-volt battery systems. The basic 12-volt output must be boosted by means of ignition coils, which are really boosting transformers.

Harleys sport two types of ignitions: electronic or point-type. Newer models employ upgraded electronic systems, which are superior to points systems. Electronic coils can be utilized in both point-type and electronic ignitions, but point-type coils can only be used in a points ignition system. When choosing coils, make sure they apply to the ignition system of your bike.

Accel, Dyna, and many other aftermarket manufacturers sell 30,000-plus voltage coils, which allow faster voltage buildup and extended spark duration. The duration of the spark at the plug needs to be long enough to provide complete combustion within the chamber.

Gains can also be realized by boosting coil output with ignition controls such as Accel's "Flame-Thrower," which extends *dwell* time and provides complete coil saturation and a voltage boost. In a points system, dwell is the time duration in which the points are closed to allow for coil power buildup, or *saturation*. In contact point, or electronic equivalent systems, point opening, not closing, initiates firing.

When using the Accel "Flame-Thrower" system you must enrich the carburetor by changing needles or jets. The higher voltage of this system provides a noticeable power increase, particularly with the newer Evolution motors. The combined use of the "Flame-Thrower" and high-voltage coils, provides an even greater increase, particularly evident in top-end power and throttle response.

The Dyna 2000 Digital Ignition System is another new, sophisticated ignition replacement system. This unit can be programmed and tailored to match the requirements of specific engine configurations. The Dyna 2000 uses advanced microprocessor technology to provide ultimate control over the complete ignition process (Fig. 4-2).

4-2
The Dyna 2000 offers programmable advance curve selection and features a built-in rpm limiter.

Two models are available and are applicable to virtually all Harley models. The RE-DP2000-HD1 is a dual fire unit for use with one dual output coil, or two dual output coils for dual-plugged engines. The RE-DP2000-HD2 is for single fire use with two single output coils, or two dual coils for dual-plugged engines.

The Dyna 2000 includes a programmable *advance curve* selection, factory optimized, to allow the ignition advance curve to be matched to the level of modification of a particular engine. A built-in rpm limiter allows you to set the appropriate limit for your engine. The limiter is adjusted by means of a four position switch on the module. A choice of 16 different rpm limits from 5,000 to 8,750 rpm is available in 250 rpm increments.

The Dyna 2000 employs a two-step method for optimum dwell control. The 2000 uses a constant dwell period to optimize spark energy, but unlike other systems, the 2000 measures charging characteristics of the coil and adjusts the dwell to match the coil's needs. The Dyna 2000 will maximize the energy output of whatever coil is used. This unit will operate with stock Hall-effect pickups or a Dyna-S as a trigger.

For the older points ignition systems, Dyna provides the RE-DBR-1 Dyna booster for single points and the RE-DBR-2 for dual points. The booster upgrades performance by transforming the usual inconsistent electrical surges into precise, controlled, bursts of energy to the coil. This lets the engine run smoother, eliminates burned points, and extends spark plug life. Condensers are eliminated, increasing sparking power for more positive combustion (Fig. 4-3).

The proper selection of spark plugs with high-intensity systems is also important. When choosing plug wires, go for the shortest wires with the lowest *Ohm* ratings. With longer wires, resistance also increases since the opportunity for current leaks is greater. The best wires are thickly insulated and the best insulation is the silicone type. Stock Harleys use 7-millimeter wires, while Sumax, Taylor, and Accel use 8- or 8.8-millimeter in their wire configurations. Insulation quality is also determined by the degree of heat protection. Harley wires are 400-plus degrees Fahrenheit; Accel 500 degrees and Sumax and Taylor wires are rated at 600 degrees.

4-3 *For points-system Harley owners, the Dyna Ignition Booster.*

Timing is also critical to electrical power transfer. Even though the spark is intensified, it must be delivered at the proper moment. The critical time of delivery is at the end of the compression stroke, just before the piston reaches Top Dead Center. Combustion must occur just after the piston reaches TDC, the position where combustion forces the piston down in the power stroke. Should timing be retarded, sparking occurs too late, causing power loss since combustion takes place after maximum compression. If timing is advanced, sparking takes place too soon before TDC and downward pressure on the piston's upward motion forces the engine to work against itself.

As the engine speed increases, the piston moves up and down faster and sparking must occur earlier, or more advanced, so that the fuel can burn fully and begin the next power stroke. The amount the timing advances as the speed of the engine increases is referred to as the advance curve. Theoretically, the longer the advance curve, the more inhibited the performance; the shorter the advance curve, the more positive the performance. In a points system, heavier points provide higher performance advance curves and curtail float at high rpm's, which in turn, eliminates misfiring or sparking at improper intervals. The newer Harleys counteract this with special ignition modules that regulate the advance curve and limit rpm's. One of the reasons the Evolution engines can accrue more mileage between engine rebuilds is because of their rev limiters.

One way to make an Evo engine run faster is to discard the stock ignition module and replace it with one from a 1200 Sportster. The installation of an Accel Mega-Fire ignition module enables you to adjust the advance curve four ways. With this unit, you can fine-tune the advance curve of a stone, stock engine through exhaust modified engines and up to radically modified motors (Fig. 4-4). The Mega-Fire unit will also allow your engine to attain higher rpm's and gain a few extra horses along the way.

There is only one optimum plug gap adjustment for a particular engine. When gapped properly, you will realize the best gas mileage and overall performance. Gaps for easy riding, light throttling, and low gas consumption are slightly wider than optimum gaps for added power and high performance.

4-4
The Accel Mega-Fire electronic control module allows you to optimize the spark timing of your Evo by selecting one of four timing curves.

Proper electrode clearance must be determined by trial and error. When a gap increase creates a reduction in performance, reduce the gap by 0.010 inch. The optimum setting is the one in which the engine performs best. Use the coldest spark plug heat range that doesn't promote fouling. The heat range of the plug might need to be changed when timing is altered or carburetors are adjusted. Richer mixtures require hotter plugs. Leaner mixtures favor colder plugs. The more efficient the ignition system, the colder the plugs that can be run.

Coils

To achieve dependable high-grade performance an ignition system must be able to generate maximum spark energy. Stock coils are O.K., but I know few true performance buffs who rely on stock Harley coils. They just don't provide the energy boost.

Dyna ignition coils are top quality, American-made coils designed for Harley-Davidson Big-Twins and Sportsters. Dyna coils produce spark voltages in excess of 30,000 volts and are intended to replace stock coils. You can choose among three styles and five variations of resistance for various applications. Dyna coils can be used with points systems, aftermarket electronic ignitions, and factory electronic ignitions (Fig. 4-5).

Another state-of-the-art American-made unit is the Accel Super Coil, for universal applications. The Accel coils are designed to provide maximum spark energy, fastest coil *rise time*, and ultra-high voltage output. Super coils mount in the

4-5
Dyna coils are capable of producing spark voltages in excess of 30,000 volts.

same manner as OEM coils. Two models are offered. The 140406 unit is for 1965 through 1979 point-fired Harley V-Twins, and the 140407 unit is for 1980 through 1993 electronically fired Twins. The finned cap acts as a voltage dam between coil outputs to prevent *arcing* and *flashover* (Fig. 4-6).

4-6
The Accel Super coil features a built-in dam to prevent arcing between wires.

For 1984 to present Evo Big-Twins, Accel offers their 35404 Super System, which provides a proper ignition component combination. Featured are the Mega-Fire ignition module, Super Coil, 8.8-millimeter plug wire set, and Accel U-groove spark plugs (Fig. 4-7).

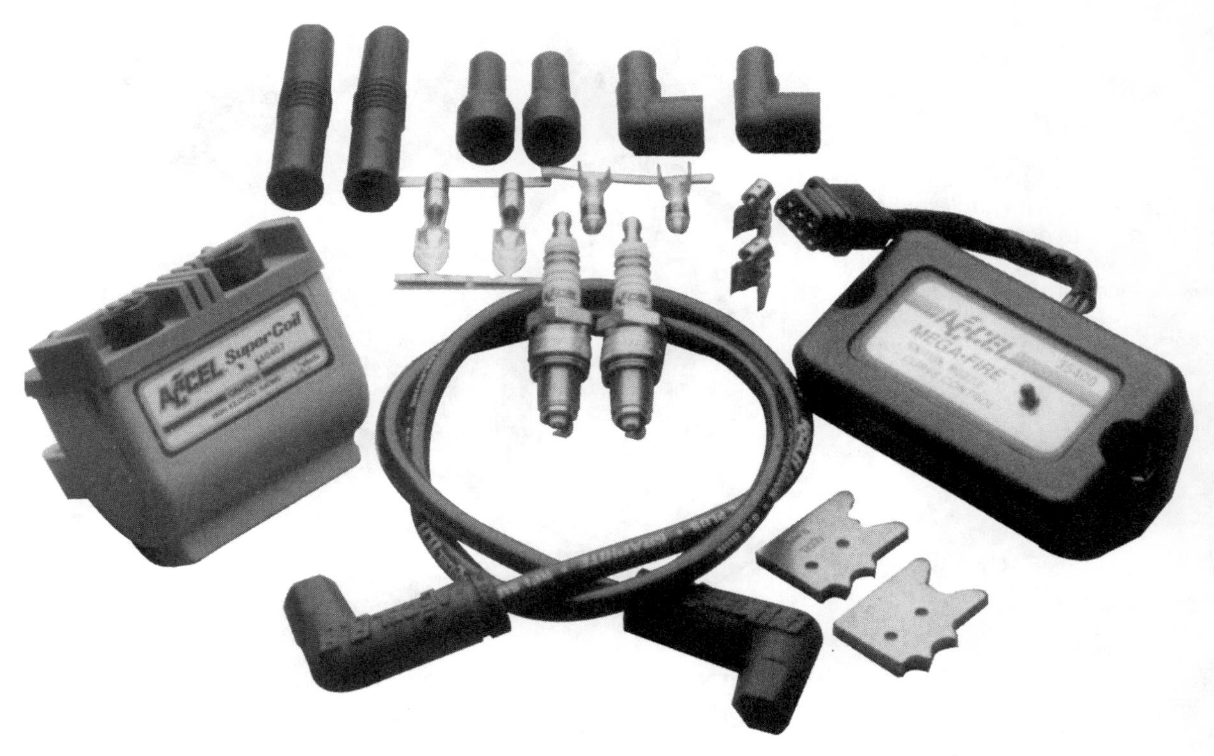

4-7 *The Accel Super System, a complete upgraded ignition system with matched high-performance computers.*

Points systems

Some riders might want to retrofit their Harleys to a dependable points ignition or even upgrade their existing points system. Accel markets a new points conversion kit complete with precision-ground breaker cam, heavy-duty breaker plate, and ultra-reliable advance weights and springs. The kit features an Accel conventional can-style condenser (Fig. 4-8). The kit provides performance upgrades for 1970 through 1978 FLs, and 1971 through 1978 FXs and XLs. All Harley Twins, 1978 to

4-8
The Accel 8505 Points Conversion Kit enables you to retrofit to dependable points ignition or upgrade an existing points system.

present, can be converted to points from electronic ignitions in a simple procedure. It is recommended that when converting to points from an electronic ignition, a high-performance coil, such as Accel's Power Pulse coil (14042) or Super Coil (14046), is used to ensure maximum spark energy. Dyna performance coils also rate high in the output department.

Single-fire ignition

Single-fire ignition systems seem to be all the rage with today's Big-Twins and Sportsters. Their advantage over dual-fire is debatable, with some folks still swearing allegiance to the conventional dual-fire system.

The traditional dual-fire system fires or sparks the plug one time for each rotation of the flywheels. With this configuration, the plug is energized not only on the power stroke, but also on the exhaust stroke, which creates a wasted spark. The extra spark firing can cause unwanted vibrations by igniting leftover fuel after the power stroke. This system has been traditional with Harley-Davidson for decades, requiring one basic coil and simple wiring. New electronic ignition systems are standard on the new Evolution motorcycles (Fig. 4-9). The newer Evos utilize a combination of the dated dual-fire system, combined with advanced electronic modules, to regulate timing.

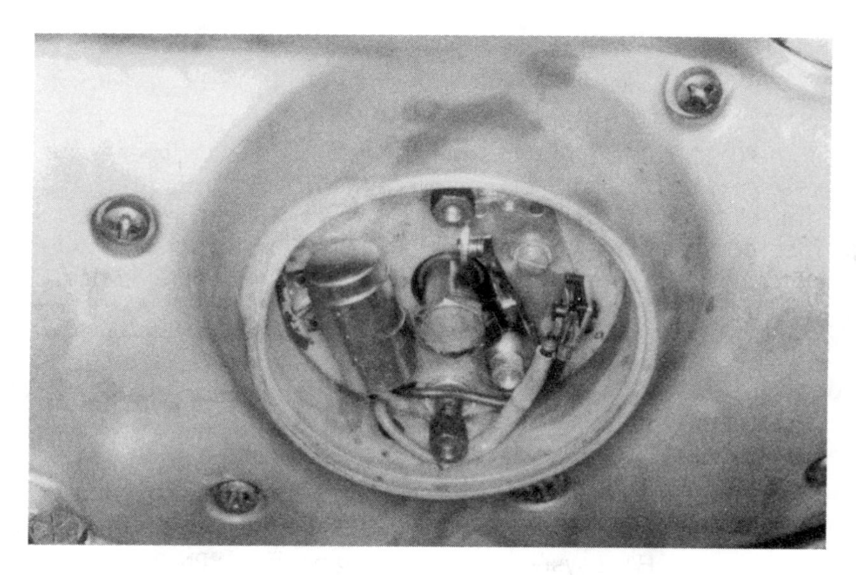

4-9
The older type points delivery system, shown here on a Sportster engine.

The newer, single-fire ignition is claimed to be more efficient and to minimize vibrations associated with a V-Twin. The choice between dual or single-fire is left to the individual; some of the better single-fire offerings are presented here.

Crane single-fire ignition

The Crane H1-2 single-fire ignition system is an excellent alternative to stock Harley ignition trigger systems. The 8-2000 system consists of two specially designed components. This magnetically triggered system contains a magnetic rotor combined with an electronic sensor plate (Fig. 4-10). Redesigned for optimum performance, the Crane H1-2 provides an additional magnet for each trigger; the magnetic triggers act as switches to control current movement to the ignition coils.

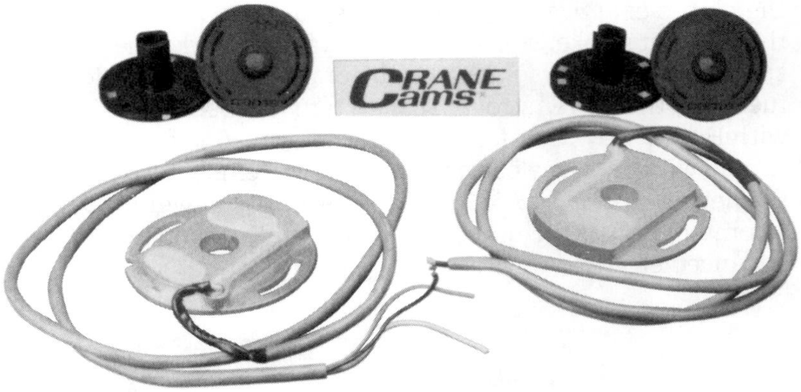

4-10
The Crane 8-2000 Hl-2 single-fire ignition and the dual-fire 8-1000 H-1 system.

Some units contain only a single magnet for each trigger so the coil remains on during engine rotation and only "off" for the time interval necessary for the coil to discharge. Crane incorporated the second magnet since the extra current flow produces extra heat at the coil, which increases coil electrical resistance. With the dual magnet set-up, the first magnet turns on the coil and the second turns it off, after energy is built up. According to Crane, this allows the system to produce accurate spark timing, providing better output from the coil to the plug.

Some newer single-fire ignitions require a mechanical advance assembly similar to the units used by Harley in the 1970 models. A highly recommended mechanical advance unit for use with the H1-2 system is Rivera Engineering's PN RECAU-1 unit. It features a stainless steel baseplate, shaft, and pins and is a highly improved version of the OEM Harley piece. In addition to the mechanical advance unit, you will need ignition coils with a 2.0 to 3.0-Ohm resistance. The H1-2 system requires two of these coils, one for each cylinder. Accel and Dyna coils serve well in single-fire systems.

For installation all you need is a test light, ring terminals for the Crane wires, some 12 to 14-gauge wire, a wire crimper, ignition wrench, and a screwdriver.

Turn on the ignition switch and probe the studs on the ignition coil to determine which one is hot. The test lamp will light when you hit the right one. Turn off the switch and remove the battery ground cable. Remove the ignition cover on the nose cone and the ignition components outlined in Crane's instruction manual, or in the Harley shop manual. When removing wires from the coil, identify each one so you know which goes to the positive and which to the negative terminal. If your model Harley has a mechanical advance, check it and replace it if it's worn for best results with a single-fire system. For 1980 and later models, follow the alignment procedure shown in Fig. 4-11. Be sure to lube the advance unit with ignition grease prior to installation.

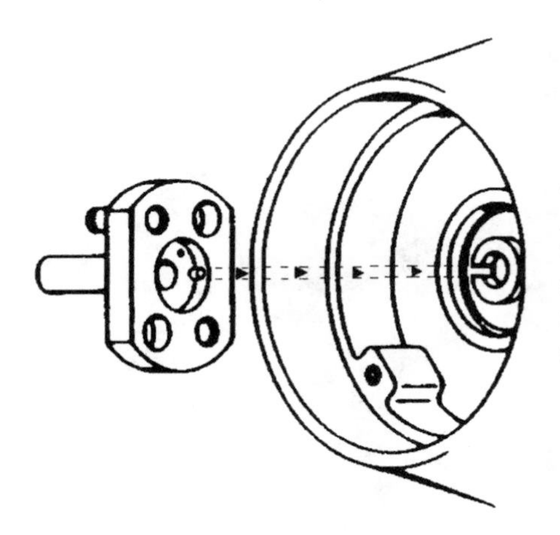

4-11
Fitting new advance mechanism to the end of the crankshaft of 1980 and later models showing the pin-alignment procedure. Crane Cams

Feed the wire from the Crane plate through the hole in the bottom of the ignition plate nose cone compartment. Then install the ignition plate. In Fig. 4-12 note the procedure and proper sequence of installation. The ignition plate has a V-notch, which should align at 7-o'clock for Big-Twins and 11-o'clock for Sportsters. Use the standoff screws that held the original timing plate in place to secure the new module. Remove and discard the circuit breaker cam, replace it with the Crane rotor (Fig. 4-13), and secure the rotor with its bolt. The rotor can go on only one way. Tighten the rotor to only 25 foot/pounds torque. Check to see that the rotor and advance assembly works freely. If the rotor interferes with the module plate, install the .030-inch washer, supplied with the kit, between the nose of the cam and the advance mechanism.

4-12
A) Standoff screws; B) Spacers; C) Rotor bolt; D) Trigger rotor; E) Module; F) Advance mechanism (not provided). Crane Cams

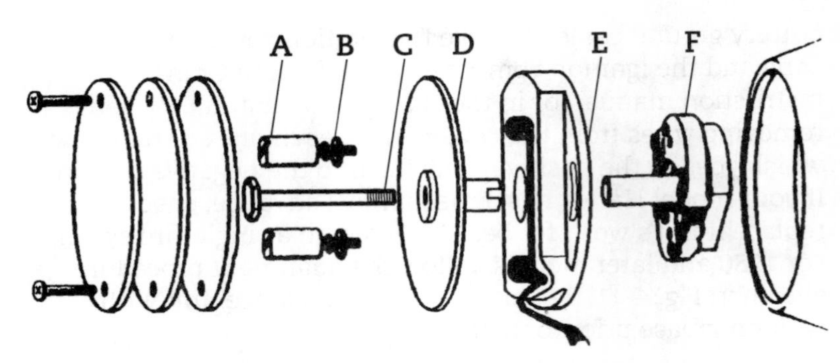

Installing the Crane hi-intensity ignition components (Part 8-2000 shown)

4-13
The rotor is installed and the rotor bolt tightened down.

Proper alignment of the advance mechanism and the nose of the cam is critical. Make sure the advance mechanism sits flat on the cam nose before tightening down the center bolt.

To finish the installation, place a ruler or straightedge across the standoff screws to determine that the head of the rotor bolt is not obstructed. If the center bolt sits too high, remove the two standoff screws and put the two small washers, provided in the kit, under them.

The H1-2 module has three wires within a protective sheath. The wires are coded black, red, and white. Figure 4-14 shows the proper procedures for both single and dual-coil applications. To obtain accurate tachometer readings with the Crane 8-2000 system, motorcycles will require a tach interface inserted in their tach cables, Crane 8-2050 tach interface module. After installing the H1-2 system, rotate the engine to the front cylinder advance mark as shown in Fig. 4-15.

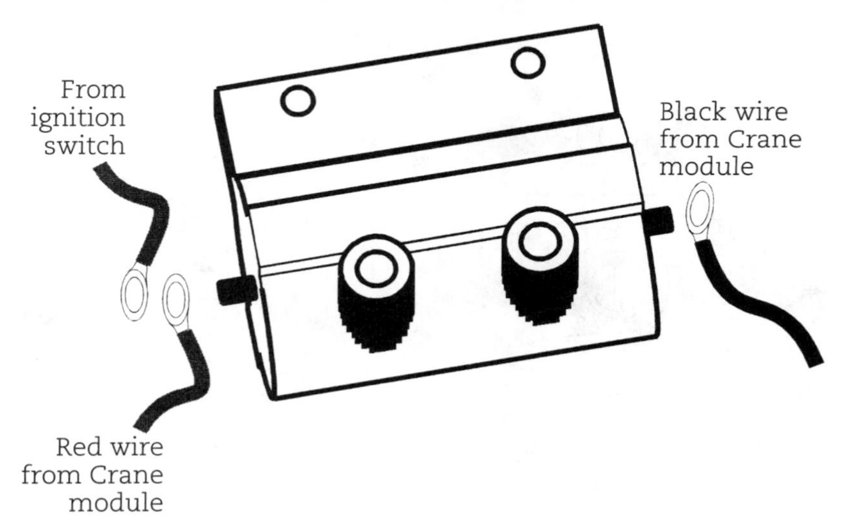

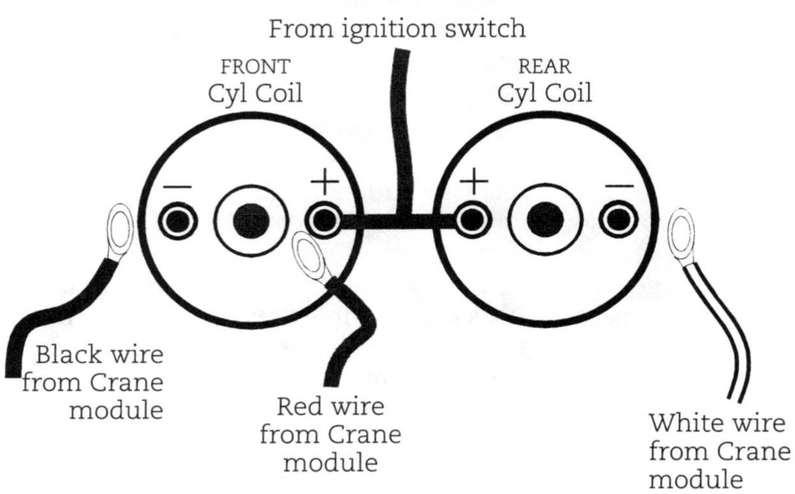

4-14
Procedure for single- and dual-coil applications.

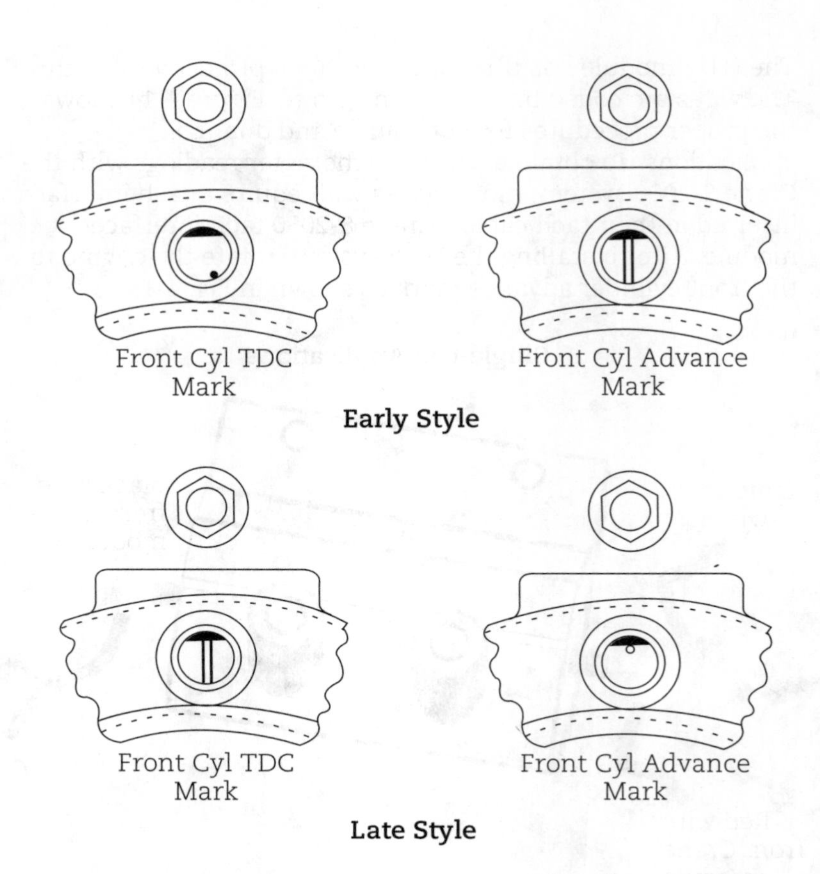

4-15
A) Rotate the engine to the front cylinder advance mark; B) Reconnect the ground cable to the negative battery terminal; C) Remove the spark plug from the front cylinder; D) Remove the plug from the timing mark observation hole; E) With the transmission in gear, rock the bike until the advance mark is in the center of the hole and the front cylinder is on compression. Crane Cams

Center the aligning mark properly and inspect the Crane ignition plate and rotor for two notches on the rotor and a raised section across the ignition plate. Loosen the standoff screws and turn the plate until the raised edge and the notches on the edge of the rotor are aligned. This will allow you to start the bike and, later, to set timing dynamically with a timing light. Make sure the standoff screws are snug, but not fully tight. This is to allow some rotation of the module plate to enable you to set dynamic timing. The procedure is known as *static timing*, or timing the engine when it is not running. Static timing is generally accurate, but you might want to confirm the accuracy because it does not take into consideration such running factors as gear lash. *Dynamic timing* is the best way to ensure timing accuracy. With plugs in place, battery and wiring

Jack Rouse's 1990 FXSTS built by Scott Barringer of Custom accessories, Pompano Beach, Fla.

Doug Doan's 1987 FXRS street racer.

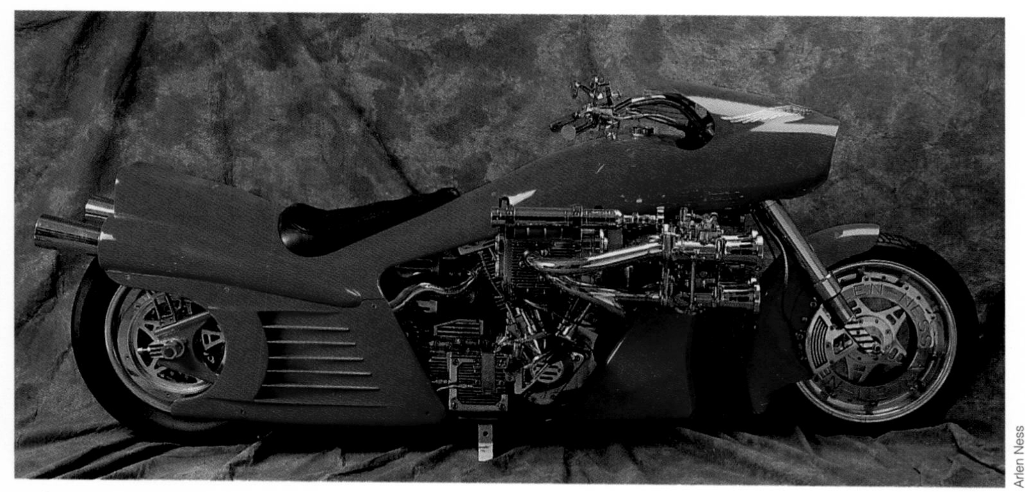

Arlen Ness

Arlen Ness's "BIG RED" $100,000 Harley Custom, with extensive hand body work, dual superchargers, four two-barrel carbs, nitrous oxide, and a 2098 cc V-twin.

Arlen Ness

Arlen Ness's "SLED." A bike within a bike, SLED features a 1987 FXR in a rubbermount softail-style frame.

Arlen Ness

Arlen Ness's 1993 Dyna Glide 80-inch Evolution.

John Sach's stroked Sportster features an air shifter and SuperTrapp exhaust.

Felix Lugo's Intimidation Sportster with a nitrous-boosted Dell'Orto dual-carb set-up.

D'Alain Ernoult & Dave Perewitz

"Vendetta." This Perewitz hand-built framed bike is owned by Paul Karing of Sharon, Pa. The bike is turbo-powered with an exhaust pipe running through the frame.

connected, and tranny in neutral, start up the engine, attach an inductive timing light to the proper battery terminals, and clip the timing light sensor wire to the front cylinder spark plug. Run the engine at about 2,000 rpm and point the light directly at the inspection hole. With each light flash you will see the correct timing mark in the inspection hole. If you need to make minor adjustments, do so by rotating the ignition module plate. Once the mark has been properly centered, tighten down the standoff screws and replace the inspection hole plug.

Dyna-S performance ignitions

The Dyna-S ignition systems have been around awhile and they are among the finest, sure-fire, and dependable. The Dyna-S units also utilize magnetic sensors, similar to the Crane system. Dyna markets two foolproof systems, a single-fire and a dual-fire type (Fig. 4-16).

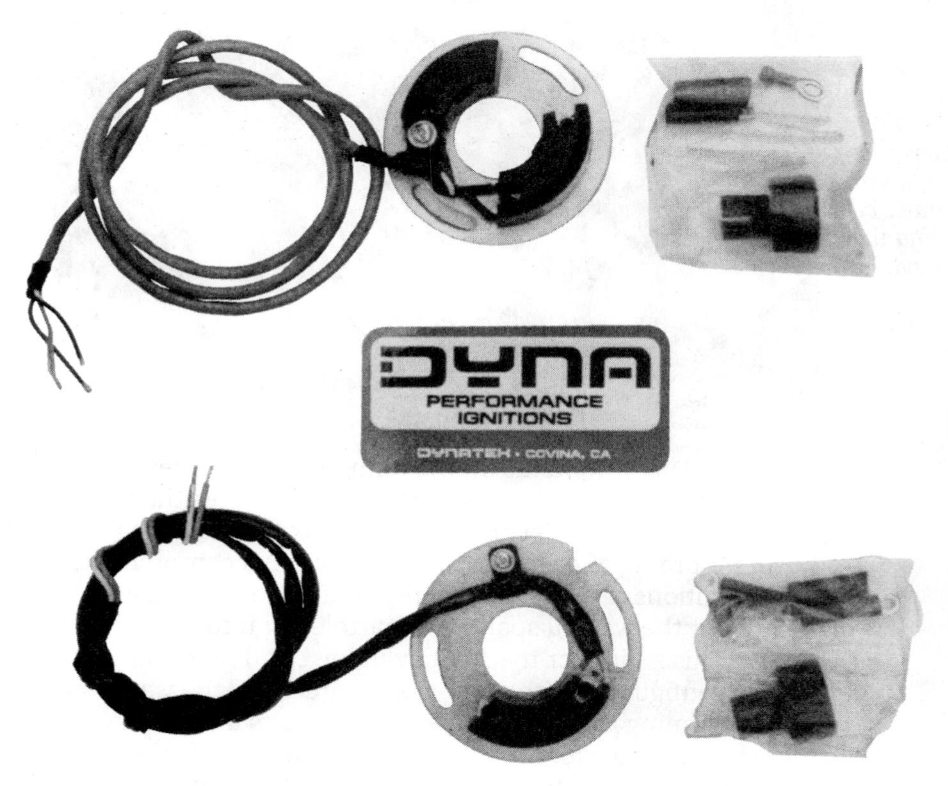

4-16
At top is the Dyna DS6-2 single-fire ignition and below the Dyna DS6-1 dual-fire unit.

The Dyna-S DS6-1 dual-fire system is self-contained and fits behind the ignition cover. It uses a magnetic rotor with the original advance mechanism so the factory advance curve on older Harley models is maintained. Later Harley models require advance units such as the Rivera RE-CAU-1 for the Dyna units to perform. The Dyna DS6-1 improves performance by varying less than 3 degrees from 0 to 12,000 rpm, half the change of some other systems. This system times faster than setting points and timing only needs to be done once, when the system is installed. The DS6-1 is ideal for dual-plug head engines (Figs. 4-17 and 4-18). In Table 4-1 and Fig. 4-19, you can see Dyna's recommended coil applications and dual plug coil set-ups.

4-17
On newer Harley Evos, the older type mechanical advance mechanism must be inserted prior to installing the electronic module.

The Dyna DS6-2 contains all the refined electronic features of the DS6-1, plus one added feature. The DS6-2 fires the front and rear cylinders independently of each other, allowing accurate timing of each cylinder rather than a compromise. Stock ignitions fire both cylinders simultaneously, the idea being that the wasted spark fires harmlessly into the exhaust stroke. This is fine for the front cylinder, but due to the 45-degree configuration, the rear cylinder receives the waste spark after beginning the intake stroke. This results in hard starting, increased engine vibration, and backfire and raw fuel spitback through the carburetor. Use of the Dyna DS6-2 single-fire ignition greatly reduces these negative factors.

4-18
The Dyna DS6-1 high performance, dual-fire module installed on a late Evolution Big-Twin.

Table 4-1 Dyna Harley-Davidson ignition/coil application

DS6-1 Ignition	**Coils**
Single plug heads (2 spark plugs)	
Race (1) 3 ohm coil	DC1-1 or DC6-1
Street (1) 5 ohm coil	DC7-1 or DC8-1
Dual plug heads (4 spark plugs)	
Race or street (2) 1.5 ohm coils wired in series	DC2-1 or DC5-1
DS6-2 Ignition	
Single plug heads (2 spark plugs)	
Race or street (2) 3 ohm coils, single output	DC3-1
Dual plug heads (4 spark plugs)	
Race (2) 3 ohm coils	DC1-1 or DC6-1
Street (2) 5 ohm coils	DC7-1 or DC8-1

4-19

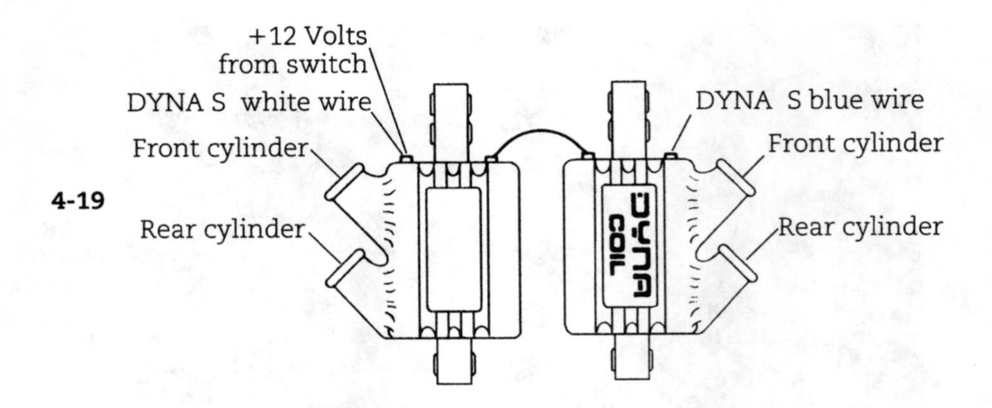

Sumax distributes an excellent solid-state single-fire ignition, manufactured by American Cycle Electronics, that aids in preventing *breakup* at high rpm's. This unit is highly efficient, providing instant spark at unlimited plug gaps and making it ideal for older kick-start Harleys. The system has built-in tach circuitry to run factory or aftermarket tachs; no adapters are needed. Dyna-S single-fire coils are recommended for use with this system as well as Sumax Taylor Spiro Pro spark plug wires. The SF-1 American Cycle Electronics system has been tested for 250,000 miles at 7,000 rpm (Fig. 4-20).

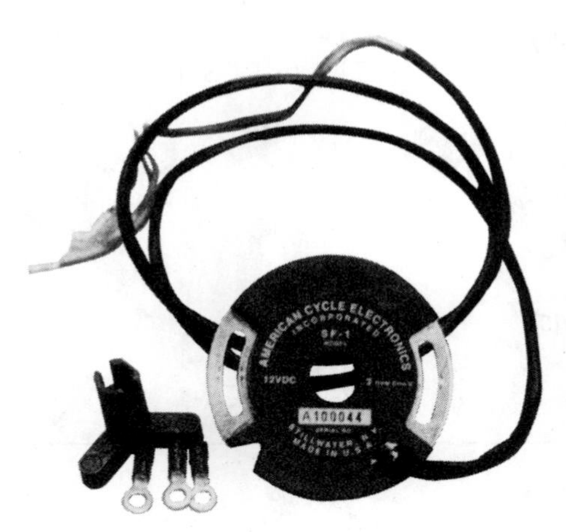

4-20
The SF-1 system, manufactured by American Cycle Electronics and available from Sumax, limits high rpm breakup at unlimited plug gaps.

Magneto ignitions

Racers and performance-oriented riders may want to consider a *magneto* ignition system. The magneto goes far back, pioneered by such companies as Joe Hunt and Morris who still make magnetos for all current and older Harleys.

A magneto ignition system doesn't rely on a battery to provide electrical power, producing its own with its own generating system. Since the power is not regulated, as with battery powered ignitions, the faster the magneto revolves the stronger and hotter the spark. Hot spark is cherished for high rpm operation, where batteries lag behind.

Magnetos, like the Morris unit shown, are costly due to the special housings that contain them and fit them to Harley engines (Figs. 4-21 and 4-22). Though they require periodic maintenance, magneto systems are the most reliable source of ignition spark. The magneto produces a positive ignition spark because the magneto reverses current polarity each time the cylinder is fired. By backing up current flow in the primary winding of the coil and promoting a discharge, the spark output is boosted considerably. The reversal of current also assists in reducing point wear. The maximum electrical discharge is delivered to the points as they open. This allows narrower point gaps, assists in eliminating *point bounce*, and curtails misfiring at high rpm's (Fig. 4-23). With a magneto system, you can realize a 137-percent voltage increase at 1,000 rpm and a 200-percent increase at 7,000 rpm. That's a lot of spark power.

When used with electric starters, magneto ignitions fire up instantly. With kick-starting you have to apply a full, long stroke, as the magneto requires at least 90 degrees of rotation of the engine to provide an adequate spark. Begin the stroke a bit past the point of the engine's compression stroke, allowing the magneto full time to fire.

Magnetos are easy to install, usually requiring replacement of the nose cone, which contains the drive gearing and timing adjustment. Morris currently supplies the finest magneto systems around, featuring models for all Harleys 1936 and on (Fig. 4-24). Units are available with manual or automatic retard.

4-21 *The Morris Magneto ignition system. The cam-driven shaft connects to the magneto housing, which contains the point cam, coil, spark plug towers, condenser, and points.*

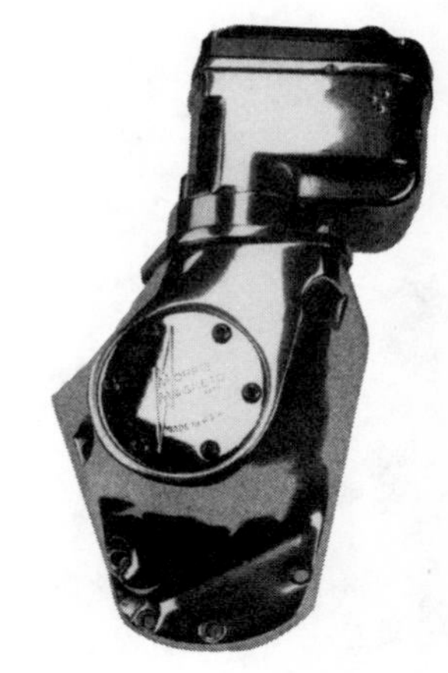

4-22 *The Morris Magneto M-5 unit with replacement nose cone for late-model 1970 and on Big-Twins. The M-5 automatically retards for starting.*

4-23
The drive mechanism of the magneto. At the bottom is the cam-driven pinion, at center the points advance system and at top the cam going into the magneto case.

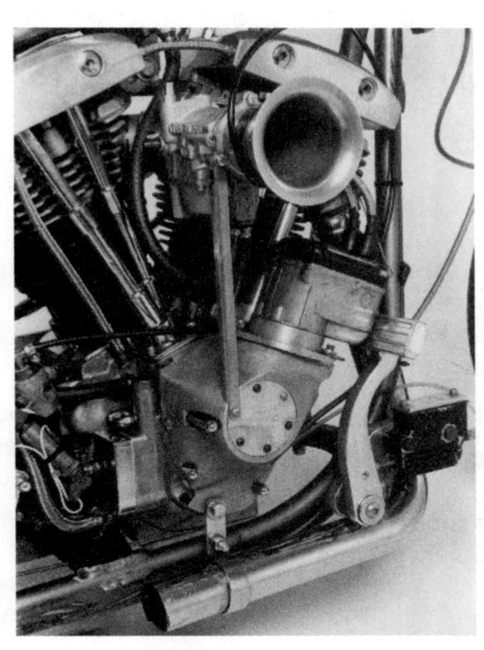

4-24
A Morris Magneto installation on Larry Stanley's "Grey Ghost," a record-winning Shovelhead dragster.

The company also offers single-fire conversion kits to be used with their magnetos. Other companies offering standard magneto systems are Karata and Joe Hunt.

Carl's Speed Shop offers a unique Capacitive Discharge Ignition magneto. This unit bolts to any Harley sidemount-style cam cover on Big-Twins 1970 and on and 1971 and on Sportsters. The kit, which requires only one hour to install, features electronic timing advance and is available for two or four plug applications.

Carburetion upgrades

Chapter 5

THE KEY TO HIGH PERFORMANCE is a well-balanced engine. One of the simplest and most important of performance add-ons is the carburetor. It is essentially a bolt-on item that does much to enhance the performance of the Harley engine. There are many available and we will consider the optimum performers as well as a few tried and true power additions that enhance carburetion.

Carburetors

The Eliminator II S.U. is one of the finest carburetors ever designed and a true performer on any Harley engine (Fig. 5-1). It is classic both in appearance and concept. The size of the *venturi* is governed by engine demand, making it highly efficient and one of the smoothest and most reliable carburetors to date.

5-1
The Eliminator II S.U. from Rivera - Primo Engineering is available with a special manifold and air cleaner and will mount to all Harley engines.

A variable venturi is created in the S.U. carburetor by the vertical motion of a close fitting piston. The piston is positioned above the main jet, centrally located in the body. Its location and shape determines the form of the upper portion of the venturi. When the piston is at rest, the venturi area is small. As the piston is raised the area increases.

Piston movement, and venturi size, is regulated by a suction disc, an integral part of the piston, located in a domed chamber housing the piston. Drilled ports in the piston transmit the pressure differential between the venturi and the throttle disc to the chamber above the suction disc. Instantly, the piston reacts to balance the pressure differential, either by raising or lowering. The piston is not cable operated but controlled only by engine demand.

As the venturi size fluctuates over wide ranges, so must the fuel orifice be varied. This is accomplished by means of a tapered needle, attached to the underside of the piston and projecting into the main jet.

Opening the cable operated throttle disc allows the low manifold pressure to be transmitted to the carburetor, as well as the chamber above the suction disc. The piston will raise, allowing the correct mixture of air and fuel to pass beneath it, relieving the low pressure. It will continue to raise until the low pressure condition has reached a value just sufficient to balance the weight of the piston (Fig. 5-2).

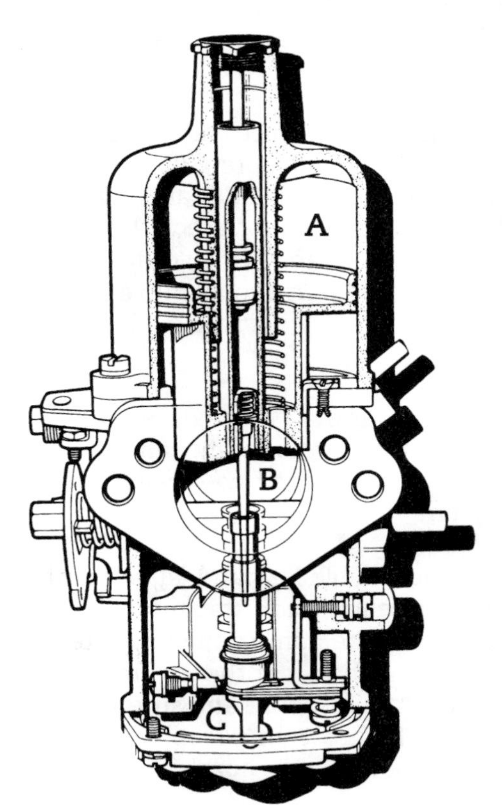

5-2
The S.U. carburetor's variable venturi is actuated by the up and down motion of the piston, controlled by vacuum in the piston chamber (A). As the piston rises, venturi size changes and the fuel orifice size varies. This is accomplished by the tapered needle, protruding from the bottom of the piston, sliding into the main jet (B). In the float bowl is a bimetallic lever, immersed in fuel (C). Changes in temperature cause the fuel jet to raise or lower and thus, regulate the mixture. Courtesy of Rivera-Prime Engineering

The Eliminator II SU carburetor has been designed to exploit the advantages gained by containing the main jet and float in a common concentric chamber, which is especially beneficial to the Harley engine. The fuel level is not seriously affected by inertial forces as the motorcycle corners, starts, stops, and encounters rough road surfaces and hills. The Eliminator II is an extremely consistent carburetor, and the concentric float adds much to its docile characteristics.

Another feature simple in design but sophisticated in function is a fuel viscosity compensating device. As fuel viscosity changes, so does the mixture. Since viscosity is directly related to fuel temperature, the float bowl contains a bimetallic lever immersed in the fuel. Small variances in temperature cause the device to act upon the fuel jet, raising or lowering it as conditions demand, thus regulating the mixture. To further provide consistency, the jet needle is held in a fixed position in the fuel jet, relative to the direction of airflow, by means of a spring loaded fixing (Fig. 5-3).

The S & S Super E is another high-performance carburetor, this one made in the U.S.A. It features a 1⅞-inch bore, fully adjustable idle mixture, and interchangeable mid- and high-range jets (Fig. 5-4). The Super E is designed for use with stock or modified 74- to 80-cubic-inch motors and is a favorite choice for stroked Evolution engines. The addition of one onto a stock 1200 Evolution Sportster or any Big-Twin will furnish admirable horsepower gains (Fig. 5-5).

The Super E flows 14 percent more air than its predecessor, the B. Features include an adjustable accelerator pump, an adjustable enrichment setting lever, and the free-flowing teardrop-style S & S air cleaner.

The E's mixture screw is easily accessible for hand adjustment. The accelerator pump's emitter stream is adjustable from zero to .3 cubic centimeters by means of the pump adjusting screw. According to S & S, the pump should be set at the minimum fuel requirement for positive throttle response and sure starting when the engine is cold. To direct the pump's emission stream to a desired position, rotate the spray nozzle with a ¼-inch wrench.

Stock Applications

Shovelhead, Sportster, Evolution

Supplied with kit as STANDARD - .100 Main Jet - Red Spring 4-1/2 oz., BBT needle.
Stock 900cc Sportster use BBD needle, 4-1/2 oz. spring, .100 Main jet.
Modified Shovelhead and Evolution engines - head work, cams, etc. Use BBT needle, .100 Main jet, silver spring 8 oz., (BBX alternative needle).
Stroked Shovelhead engines to 93 cu. in. Use .1015 Main Jet, BBT needle, 8 oz. silver spring.
Stroked Evolution engines to 93 cu. in. Use BBT needle, .1015 Main jet, 8 oz. spring, BBX alternate needle.
All needles for both Eliminator I & II are supplied with collar attached to shank for use as spring loaded needles.Needles: LEANEST to RICHEST - BBD, BBT, BBX, BCJ, BBZ. BBT standard.
For STROKED engines the main jet should be changed to Rivera Part No. RE-1453A (.1015) or RE-1453B (.1024).
All Eliminator SU Carburetors supplied by Rivera Engineering for stock or near stock applications are supplied with a BBT needle, .100 main jet, and a 4-1/2 oz. spring. Spare jetting includes a richer needle (BBX) and a leaner needle (BBD) or substitutes.
The main jet is adjusted for starting purposes when shipped. If it ever becomes necessary during tuning procedures to turn the main jet adjustment screw more than two full turns either way, a different tapered needle should be installed and the main jet set to the neutral position for further tuning.
The tapered needle is fixed inside the piston. NEVER loosen the set-screw and raise or lower the needle. The needle guide which holds the needle should be flush with the bottom of the piston.
Standard installed needle and seats are of the GROSE-JET type, designed for gravity flow. For racing purposes or large stroker engines we suggest using our fuel bowl spacer, Part No. RE-680-S. This part will double the float bowl capacity.
Our PRIMER PUMP (Pat. No. 4,228,110) is a pressure system. The pump is a press fit inside the body and SHOULD NOT BE REMOVED. The brass nut can be removed to clean the inside or change the viton cup if necessary.
If FUEL should drip or flow from the tickler pump, shut off the fuel immediately. The bottom cover will have to be removed to visually check the GROSE JET or the fuel level. It is possible dirt will cause the ball to stick inside the seat. It can be removed for cleaning or lightly blown out with an air hose.
The float level should also be checked and adjusted as shown in SETTING THE FLOAT LEVEL illustration.
Piston springs are a tuning asset. Stock carburetors have the weakest spring installed (4-1/2 oz.). By changing springs we can accommodate a slight mixture change. A stronger spring will richen the mixture over the entire RPM range.
NEVER use any oil inside the dampner. Oil will slow the rise of the piston causing an overly rich mixture. Every 30 days unscrew the dome cap and lightly spray around the piston shaft with WD-40. The piston must float freely at all times to accomplish the constant velocity principle of allowing the engine to determine its position.
Rivera Engineering manufactures stroker size manifolds for popular stroker kits. If you require a stroker manifold please ask when placing carburetor orders.
The Eliminator II SU is manufactured by the SU factory to Rivera Engineering blueprints. Each carburetor is completely disassembled, modified and reassembled before sold. QUALITY CONTROL is assured.
All Chrome or Half Chrome carburetor kits are supplied without a rod on the dampner cap. As we manufacture this dome cap and use no oil in the dampner a rod is not necessary. If for some reason you require a complete dampner we will exchange the chrome cap for a plastic cap with a rod.

ELIMINATOR SU NEEDLE GUIDE

LEANER → RICHER

BBD	BBT (Standard)	BBX	BCJ	BBZ
.099	.099	.099	.0995	.098
.095	.096	.095	.0967	.0954
.092	.0932	.0932	.0939	.0924
.090	.0903	.0905	.0909	.0892
.088	.0877	.0875	.0881	.0862
.0862	.0850	.0852	.0848	.0819
.0844	.0827	.0829	.0781	.0780
.0825	.0807	.0806	.0740	.0751
.0818	.0792	.0782	.0703	.0713
.0808	.0778	.0755	.0671	.0673
.0798	.0765	.0730	.0650	.0653
.0788	.0753	.0702	.0630	.0629
.0778	.0740	.0675	.0610	.0605
.0768	.0725	.0650	.0590	.0580
.0758	.0713	.0624	.0570	.0560
.0748	.0700	.0598	.0560	.0540

NOTE:

The numbers under each needle is the profile dimension of that needle.

Dimensions taken every 1/8" from top shoulder.

5-3 *S.U. needle and jet guide, these specs apply to all Shovelhead, Sportster, and Evolution engines.*

Super E and G Carburetor Body and Related Parts

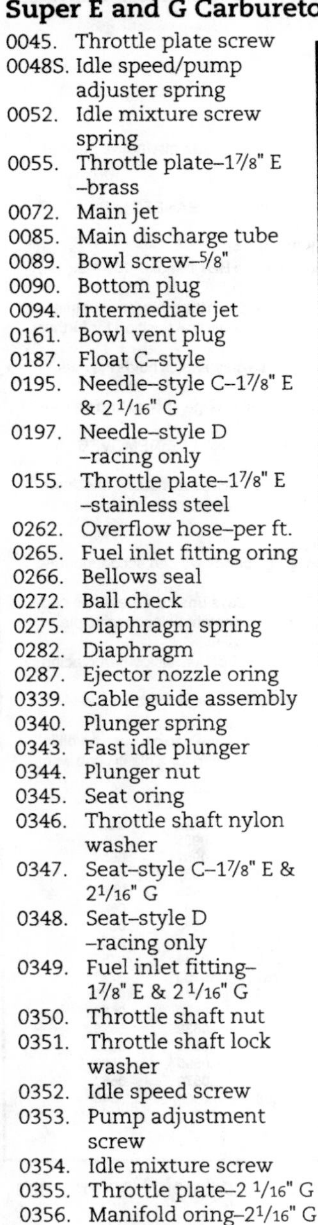

0045. Throttle plate screw
0048S. Idle speed/pump adjuster spring
0052. Idle mixture screw spring
0055. Throttle plate–1 7/8" E –brass
0072. Main jet
0085. Main discharge tube
0089. Bowl screw–5/8"
0090. Bottom plug
0094. Intermediate jet
0161. Bowl vent plug
0187. Float C–style
0195. Needle–style C–1 7/8" E & 2 1/16" G
0197. Needle–style D –racing only
0155. Throttle plate–1 7/8" E –stainless steel
0262. Overflow hose–per ft.
0265. Fuel inlet fitting oring
0266. Bellows seal
0272. Ball check
0275. Diaphragm spring
0282. Diaphragm
0287. Ejector nozzle oring
0339. Cable guide assembly
0340. Plunger spring
0343. Fast idle plunger
0344. Plunger nut
0345. Seat oring
0346. Throttle shaft nylon washer
0347. Seat–style C–1 7/8" E & 2 1/16" G
0348. Seat–style D –racing only
0349. Fuel inlet fitting–1 7/8" E & 2 1/16" G
0350. Throttle shaft nut
0351. Throttle shaft lock washer
0352. Idle speed screw
0353. Pump adjustment screw
0354. Idle mixture screw
0355. Throttle plate–2 1/16" G
0356. Manifold oring–2 1/16" G
0368. Bowl screw–2 7/16"
0369. Float pin
0371. Pump cap oring
0373. Pump pushrod
0374. Ball check spring

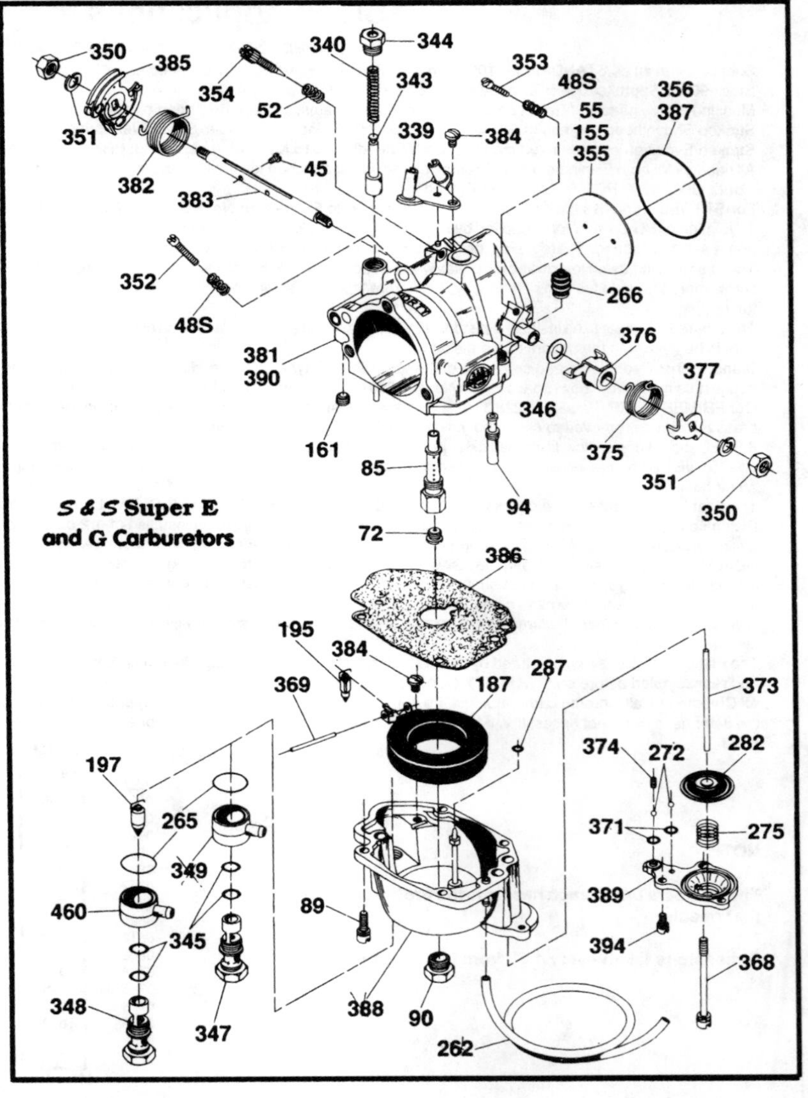

0375. Actuator spring
0376. Pump actuator lever
0377. Pump actuator arm
0381. Carb body assembly–1 7/8" E
0382. Throttle return spring
0383. Throttle shaft
0384. Cable damp/float pin screw
0385. Throttle spool
0386. Bowl gasket
0387. Manifold oring –1 7/8" E
0388. Carb bowl assembly
0389. Pump cap assembly
0390. Carb body assemby–2 1/16" G
0394. Pump cap screw
0460. Fuel inlet fitting–racing only

5-4 *Super E carburetor body and related parts.*

5-5
The S&S Super E "Shorty", a 1⅞-inch bore butterfly-type carburetor with adjustable idle mixture and replaceable jets. A is the idle mixture screw. B is the accelerator pump screw.

The Super E's enrichment lever is located behind the air cleaner, easily accessible, which is essential for regulation of mixtures over a wide range (Fig. 5-6).

This S & S carburetor is the heart of Carl Morrow's Git Kit, which also includes Carl's True Flow matched exhaust system, intake manifold, special performance cam, and special pushrods. The kit is available for Shovelheads, Evo Big-Twins, and 1986 through 1991 Evo XLs (Fig. 5-7).

The Dell'Orto "Pumper" is an excellent Italian single-throat carburetor and an ideal high-performance replacement part for stock Harley engines. It has been widely accepted, particularly in racing applications. Available from Rivera Engineering, this refined carburetor is available in 38- or 40-millimeter bore sizes and will fit all Evo engine models and retrofit to all pre-Evolution Harleys. The "Pumper" features a built-in accelerator

5-6
The venturi and throat of the S&S Super "E". S&S is known for their high-efficiency carburetor designs.

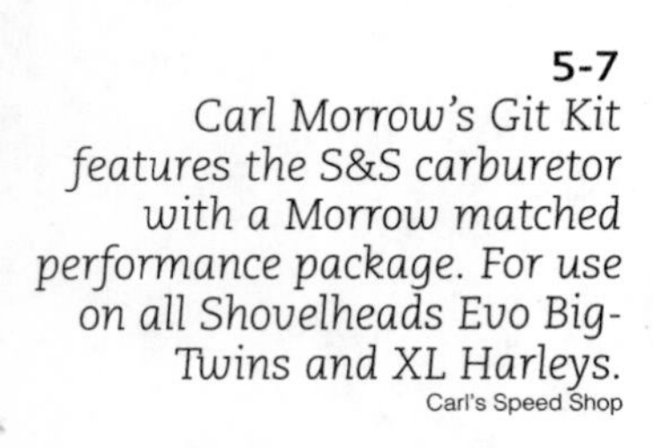

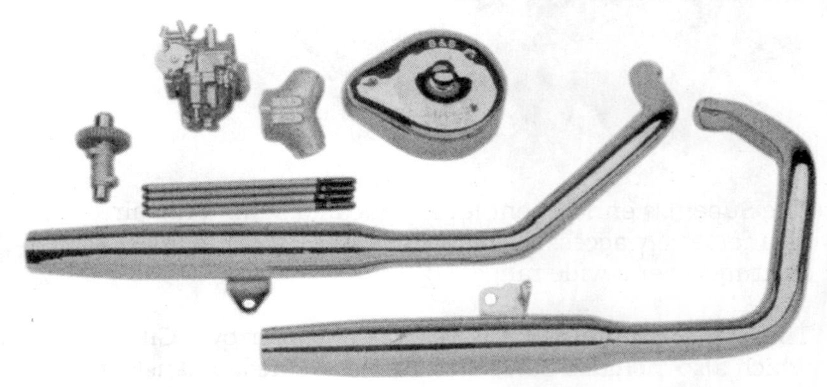

5-7
Carl Morrow's Git Kit features the S&S carburetor with a Morrow matched performance package. For use on all Shovelheads Evo Big-Twins and XL Harleys.
Carl's Speed Shop

pump, variable jets, and adapters and air cleaners for virtually all applications (Fig. 5-8).

There is also available a dual "Desmo" style pumper set-up, coupled with a new Evolution manifold designed by Jerry Magnuson of Fueling Engineering. This system is specially designed for use with the Fueling-Rivera four-valve heads and is progressively actuated. After 30 percent of the primary carb is engaged, the secondary carb comes into play. The linkage is set up so that when the secondary carburetor engages, the two

5-8
The Dell'Orto "Pumper" is available in 38- and 40-millimeter throat sizes, popular with many motorcycle racers because of their performance and dependability.

work together to reach full throttle simultaneously. This carburetion set-up is available for stage II motors using the Fueling-Rivera four-valve heads (Fig. 5-9).

Perfectly suited for all Harley applications, from stocker to stroker, is the 40-millimeter Dell'Orto Dual Throat (Fig. 5-10). This is a side-draft type carburetor and contains two identical induction barrels controlled by dual synchronized throttles. Each of the barrels provides fuel independently to a single engine cylinder. All tuning can be performed without removing the carburetor from the manifold. Extra jetting and complete tuning instructions are included with each kit available from Rivera Engineering (Fig. 5-11). Carburetors can be custom-jetted according to customer specifications or can be obtained pre-jetted for stock Big-Twin and Sportster engines.

5-9
The Dell'Orto Desmo System, progressively linked by means of a slider bracket on the right side and a connecting linkage to the throttle on the second carb.

5-10
A dual Dell'Orto mounted on a Sportster. Special manifolds are available for early model Shovelheads and Sportsters.

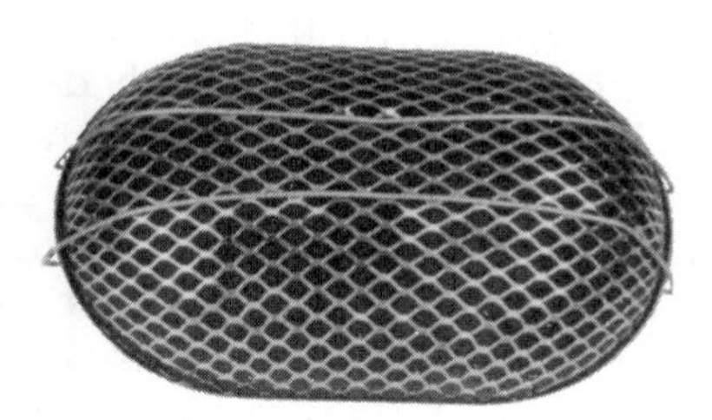

5-11
The Dual Dell'Orto is marketed by Rivera Engineering. Air cleaners as shown are available as well as the can type and velocity stacks.

One of the most important features of this Dell'Orto is *air bleed* correction, which is governed by the *emulsion* tubes, two specialized jets that work in conjunction with various other jets and a spray tube. Over a dozen emulsion tubes and idle speed jets can be interchanged to provide infinite fuel metering and optimum performance. The purpose of the emulsion tubes is to emulsify metered air emanating from the bleed jet with fuel issuing from the main jet. The tube's influence is more apparent at small to medium throttle openings and throughout acceleration (Fig. 5-12). The recommended combination for

5-12
Easily interchangeable, the Dell'Orto's emulsion tubes offer a wide range of tuning. Emulsion tubes and idle jets are located in a common chamber.

stock Shovelhead and Evolution engines is a #1 idle jet and #5 emulsion tube. The distinct advantage of the dual Dell'Orto's corrective bleeding is improved fuel atomization.

The Qwik-Silver II, a recent entry in the performance carburetor race, is a newly designed, American-made unit distributed and sold by Rivera Engineering. Designed by Red Edmonston, who created the Lake Injector, Posa Fuel, Lectron, and Blue Magnum fuel systems, the Qwik-Silver II incorporates a new slide mechanism that regulates fuel flow and needle operation (Fig. 5-13). Qwik-Silver II custom carburetors come pre-jetted for a particular model. There are no needles, jets, or slides to adjust or replace. Theoretically, it's a bolt-on and go unit. The carb contains an internal compensation system that automatically adjusts for rpm, altitude, and weather changes (Figs. 5-14 and 5-15).

5-13
The Qwik-Silver II carburetor in side view. Presetting eliminates the need for outer controls.

5-14
Special manifolds are marketed by Rivera for various Evo and Shovelhead applications, in complete kits as shown.

To be released in late 1993, the new Mikuni HSR-42 with 42-millimeter venturi flows 20-percent more air than its predecessor, the HS-40. It is particularly applicable to the 1340 Evolution engine. The HSR-42 does not rely on a conventional butterfly valve and can flow more air at full throttle to produce more peak power.

The new Mikuni features a larger capacity float bowl and a freer flowing needle valve assembly to enhance extended full throttle operation. A high leverage float valve prevents float bowl overfill and needle valve flooding, eliminating the need for overflow and vent tubes. The unit also contains a built-in accelerator pump, adjustable to produce instant throttle response through the rpm range.

Replacing the conventional butterfly valve is a roller bearing throttle slide, which ensures smooth slide action with less throttle pull. The flat throttle slide mechanism provides an unobstructed venturi at full throttle. The HSR-42 throttle pull is much lighter than its predecessor the HS-40 because of the roller bearing slide and a light throttle return spring. The HSR-42 will initially be available for late-model Evolutions, Shovelheads, and late Sportsters.

PARTS LIST

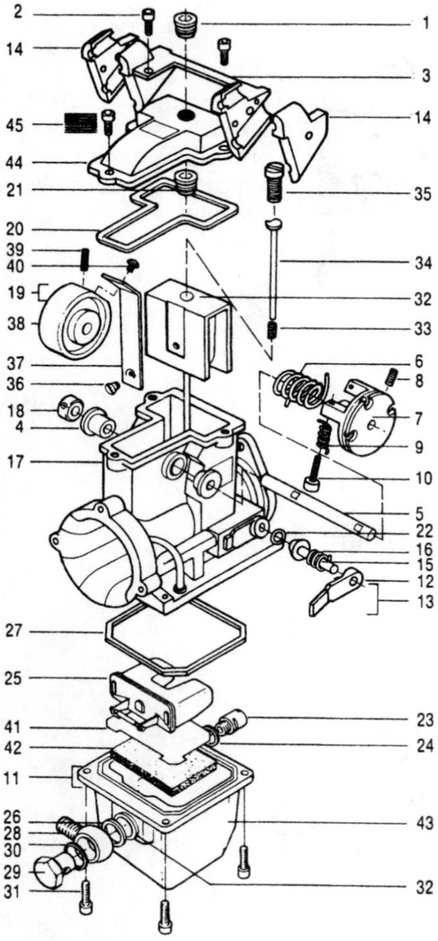

ITEM NUMBER	NUMBER REQUIRED	PART NUMBER	DESCRIPTION
		222-0000	**HARLEY DAVIDSON** CARBURETOR COMPLETE 36mm; 38mm; 40mm; 42mm **INDICATE YEAR MODEL & ALL ENGINE MODIFICATIONS**
1	1	222-0001	1/4 NPT PLUG
2	7	222-0002	#8 SOCKET HEAD CAP SCREW
3	1	222-0003	BRACKET
4	2	222-0004	BEARING
5	1	222-0005	SHAFT
6	1	222-0006	SPRING INDICATE LEFT OR RIGHT WINDING
7	1	222-0007	CABLE WHEEL
8	1	222-0008	SET SCREW #8
9	1	222-0009	IDLE TENSION SPRING
10	1	222-0010	ALLEN HEAD IDLE SCREW
11	1	222-0011	FLOAT BOWL ASSEMBLY COMPLETE
12	1	222-0012	CHOKE LEVER
13	1	222-0013	CHOKE COMPLETE
14	1	222-0014	CABLE ATTATCHMENT
15	1	222-0015	CHOKE PIN
16	1	222-0016	CHOKE SPRING
17	1	222-0017	BODY ASSEMBLY **INDICATE YEAR MODEL & ALL ENGINE MODIFICATIONS**
18	1	222-0018	SET COLLAR
19	1	222-0019	SLIDE ASSEMBLY COMPLETE
20	1	222-0020	GASKET
21	1	222-0021	NEEDLE ADJUSTMENT SCREW
22	1	222-0022	CHOKE "O" RING
23	1	222-0023	SEAT
24	1	222-0024	SEAL
25	1	222-0025	FLOAT ASSEMBLY
26	1	222-0026	GASKET
27	1	222-0027	GASKET
28	1	222-0028	BANJO INDICATE SINGLE OR DOUBLE
29	1	222-0029	BANJO BOLT
30	1	222-0030	GASKET
31	1	222-0031	SCREW
32	1	222-0032	SLIDE
33	1	222-0033	SPRING
34	1	222-0034	NEEDLE LEAN 3 thru RICH 18 **INDICATE YEAR MODEL & ALL ENGINE MODIFICATIONS**
35	1	222-0035	ADJUSTMENT SCREW
36	1	222-0036	SCREW
37	1	222-0037	ACTUATOR STRIP
38	1	222-0038	SPOOL
39	1	222-0039	SET SCREW
40	1	222-0040	PHILLIPS SCREW #4
41	1	222-0041	RETAINER PLATE
42	1	222-0042	FUEL CELL FOAM
43	1	222-0043	FLOAT BOWL
44	1	222-0044	CAP
45	1	222-0045	AMERICAN FLAG DECAL
46	1	222-0046	TUBE

5-15 *Exploded view of the Qwik-Silver II carburetor.*

For those who do not want to go the new-carb route, there is an excellent alternative that will give your existing carburetor more power. It's called the "Thunderjet" and is available from Zipper's Performance Products or your favorite Harley shop. Performance parts on Harleys put more demands on the carburetor and the Thunderjet helps meet these demands (Fig. 5-16).

5-16
The Thunderjet, showing the additional fuel nozzle. This adds an additional fuel circuit, augmenting overall power.

Most carburetors have basically two fuel circuits. At low rpm's, the low or midrange circuits can be modified and re-jetted to provide a proper mixture. Jetting can also solve mixture problems in the high speed or main jet circuit. The compromise is in the area between low and high range circuits, resulting in flat spots with mixture *leanout*, or ragged running produced by too rich a mixture. The Thunderjet bridges this gap by adding on additional fuel circuit. Smooth, uninterrupted power can be realized even with super carbs like the S&S and super-stroked engines.

The Thunderjet is not a power jet. A power jet has no indicative atmospheric air bleed. Fuel flow is controlled by the vacuum incurred in the carburetor venturi. A power jet is capable of feeding fuel into a motor running lean under specific conditions, but unable to ascertain subtle variations in airflow. In comparison, the Thunderjet responds to a momentary fuel demand by instantly reacting to pulses in the intake tract, when a motor leans out. The Thunderjet responds by allowing more fuel to flow. The Thunderjet has been proven to dramatically improve throttle response in such carburetors as the S&S Super G and Super E, Bendix, Rev-Tech, Keihin, and similar carburetors.

Installation is straightforward and does not require special tooling, just a drill and a tap as indicated in the instruction sheet. After a hole is drilled in the throttle body, in close proximity to the low speed jet, it is tapped and the Thunderjet fuel-feed nozzle is screwed into the carburetor body (Fig. 5-17).

5-17
After the carb body is drilled and the hole tapped, the Thunderjet nozzle is screwed in.

Optional jets, with varied fuel flow characteristics to widen the Thunderjet's range, are also available (Fig. 5-18).

5-18
Auxiliary jets for the Thunderjet are available and screw in easily.

A second fitting is installed in the float bowl, connecting to the Thunderjet by means of a plastic feeder hose, to draw out the fuel from the float chamber (Fig. 5-19).

5-19
The Thunderjet installed on a stock Keihin carburetor. The nozzle is fed by the tube attached to the fitting in the bottom of the float bowl.

Proper fuel flow is crucial to high performance. Pingel high-flow fuel valves are recommended to replace stock valves in the Harley tanks, Fatbobs or otherwise. As an added precaution, use Pingel free flow in-line filters between the gas tank and carburetor. Dirt free fuel is essential to optimum carburetor functioning (Fig. 5-20).

5-20
The Pingel fuel filter is a wise addition.

Manifolds

The manifold might seem a mundane commodity but it is essential to good flow between the carburetor and the heads. A good, well-designed manifold can boost an engine's horsepower and improve gas mileage, whereas, a poorly designed, malfunctioning, or leaking manifold can hinder performance. Mixture flow must be directed and the job of the manifold is to direct the flow equally and efficiently to each cylinder head. Many of the newly marketed manifolds solve the problem of the 1984 to 1989 Evo's cracking of the rubber manifold fittings.

Some of the newer manifolds such as Ram Jet and S&S have improved flow characteristics and these units will retrofit back to 1984 models (Figs. 5-21, 5-22, and 5-23).

5-21
The S&S manifold, designed for S&S carbs, is an excellently flowing manifold for Evo engines and is available for Big-Twins and Sportsters.

5-22
A wide selection of special manifolds for Panhead, Shovelhead, and XL motors is available from Custom Chrome.

Arlen Ness markets a replacement kit for the rubber compliance fitting, utilizing the stock Evo manifold. The kit is designed to use an O-ring in a groove for sealing when the stock manifold is seated inside the heads. Stock manifold supports are used. A thicker manifold-to-carburetor gasket is also supplied.

5-23
A remedy for damaged compliance fittings on Evolution Big-Twins, and XLs, 1984 and on, is the Ram Jet intake manifold by Custom Chrome.

Air cleaners

Air cleaners are essential to good breathing and there are many aftermarket accessory types that really augment airflow. Not only do they work well but are attractive to boot, offering a custom flair to go with good airflow (Fig. 5-24). Arlen Ness offers billet aluminum air cleaners. These innovative air cleaners will fit on most 1984 and later stock Harley carbs, and on such carbs as the S&S with the standard 3 bolt mounting pattern.

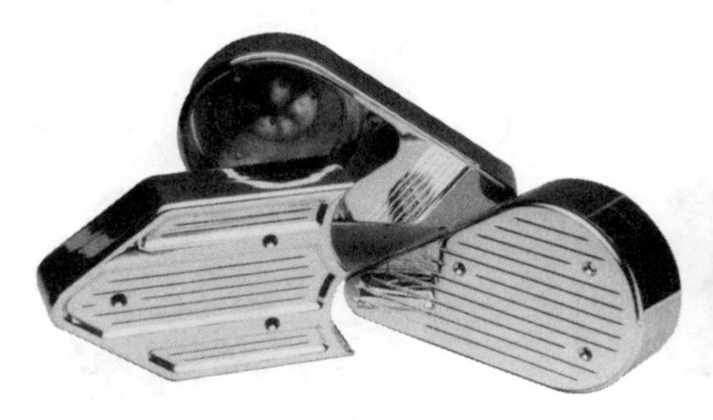

5-24
These lustrous air cleaners are from Arlen Ness. They are high-grade billet aluminum and the best looking units around.

Unquestionably the most unique, functional, and eye-catching air cleaner is the "Hypercharger," patented and marketed by Küryakyn. Fully automatic, the Hypercharger's vacuum pod opens the butterflies as the motor requires more air. The unit utilizes a replacement K&N filter element and is purported to outflow all current accessory or stock air cleaners. The Hypercharger is a performer on all engines up to 160 cubic inches. It will easily mount to 1966 to 1994 Harleys and on most aftermarket carbs, with the proper adapters (Fig. 5-25).

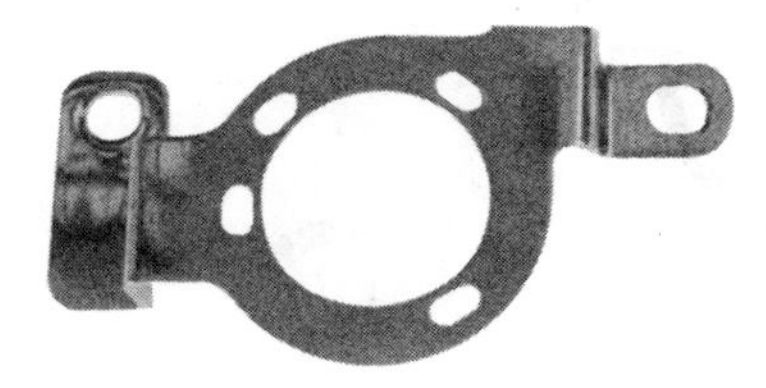

5-25
The Hypercharger by Küryakyn. With progressive intake control, this unit claims to outflow all other aftermarket air cleaners.

Prior to mounting, reroute the stock vacuum hose located behind the carburetor, under the fuel tank, to the T-fitting as shown in Fig. 5-26. Attach the accessory mounting bracket to the cylinder heads using the two buttonhead bracket mounting bolts. Install the Hypercharger using the special gasket, which mounts between the cylinder head bracket and the air cleaner housing. The Hypercharger can be rotated to any desired angle but looks and works best when level. Remove the crankcase vent tube, as the Hypercharger doesn't utilize crankcase ventilation. Replace the vent tube with an aftermarket crankcase ventilation breather.

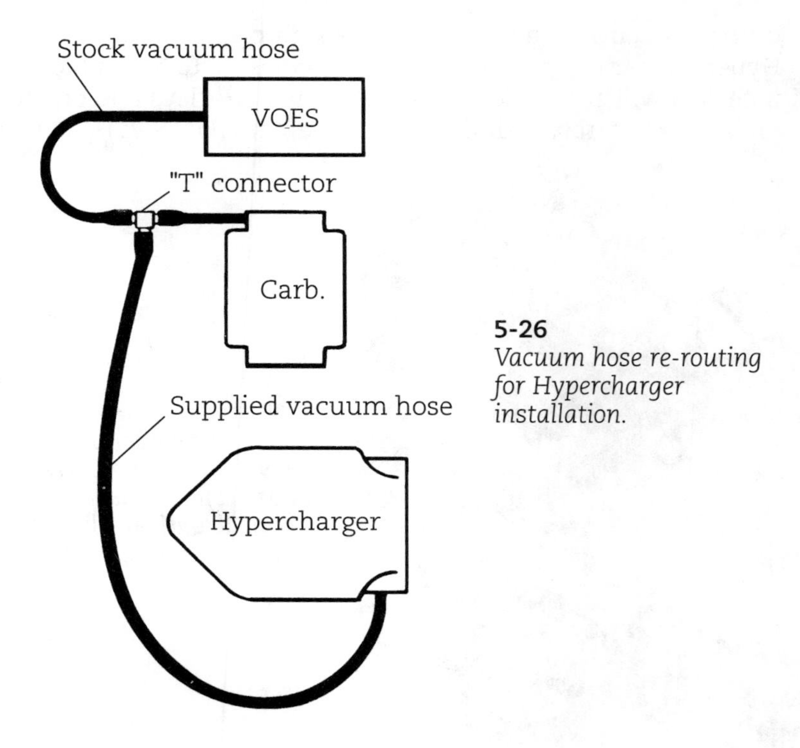

5-26
Vacuum hose re-routing for Hypercharger installation.

The normal position of the butterflies on the intake port is open when the engine is not running, and almost closed at idle. The butterflies open as the throttle opens and are factory adjusted for full open to closed positions and require no further adjustment (see Fig. 5-27 and Table 5-1).

Hyperchargers are available for Evo Big-Twins and pre-Evolution Shovelheads with the same bolt pattern as Keihin carbs. They are also available for 1991 to 1993 Sportster XLs.

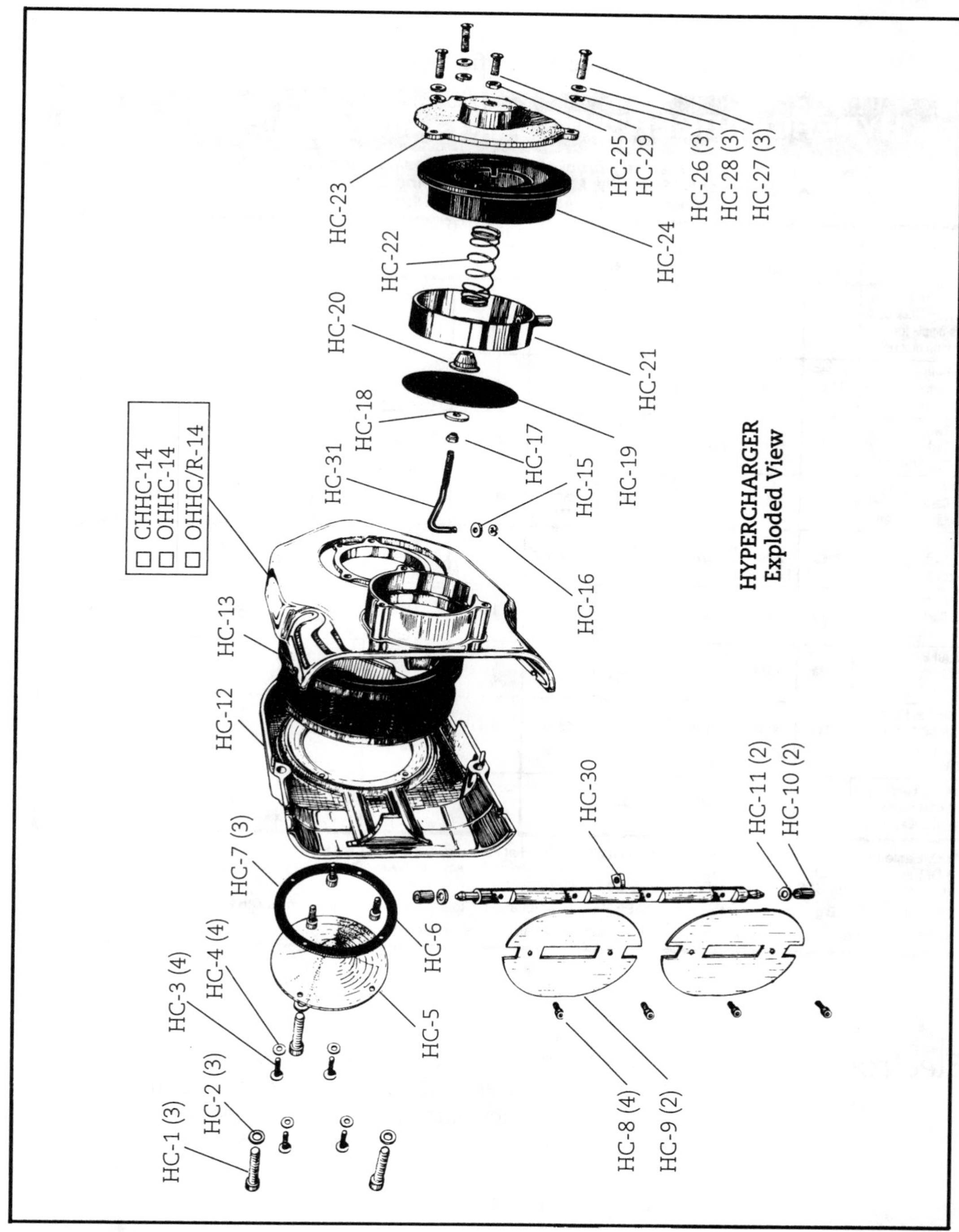

5-27 *Exploded view of the Hypercharger.* Küryakin

Table 5-1 Hypercharger fitments and adapters.

HYPERCHARGER™ FITMENTS AND ADAPTERS

Air Cleaner includes all necessary brackets, bolts and gaskets to mount to stock carb. Adapters for S&S Shorty, Mikuni and Screaming Eagle are extra. Crankcase breather/filter units are also extra.

	RED	BLACK	BLUE	TURQUOISE	E&G/S&S ADAPTER	HS40 MIKUNI ADAPTER	SCREAMING EAGLE	CRANKCASE BREATHER	S&S ENRICHENER	MOUNTING KIT ONLY	REPL. FILTER ELEMENT
66 & Later – Pre-Evo Big Twin & XL f/Stock Carb Includes nipple for manifold; no bracket.	8450	8451	8452								8513
84-89 Big Twin f/Stock Carb No adapter necessary	8495	8496	8497							8335	8513
90-91 Big Twin f/Stock Carb No adapter necessary	8500	8501	8502	8503	8520 ***	8521	8522	8515	8519	8336	8513
92 Big Twin f/Stock Carb No adapter necessary	8507	8505	8508	8509	8520 ***	8521	8522	8515	8519	8337	8513
93 Big Twin f/Stock Carb *See note below.	8455	8456	8457		8535 ***	Available Soon	Available Soon		8519	8340	8513
88-90 XL All Models f/Stock Carb No adapter necessary	8355	8356	8357	8358	8520 ***	8521	8522	8515	8519	8338	8513
91-93 XL All Models f/Stock Carb **See note below.	8360	8361	8362	8363						8339	8513
Pre-Evo Big Twin & XL w/S&S/E&G Shorty Includes nipple for manifold; no bracket.	8447	8448	8449		8520 ***	8521	8522	8515	8519		8513
84-91 Evo Big Twin w/S&S/E&G Shorty, Mikuni, Scr. Eagle	8500	8501	8502	8503	8520 ***	8521	8522	8515	8519		8513

* H.D. Eagle Iron #29308-93 Crankcase Breather must also be used at this time. KüryAkyn™ will also be offering a kit soon.
** H.D. Eagle Iron #29281-91T Crankcase Breather must also be used at this time. KüryAkyn™ will also be offering a kit soon.
*** Be sure to order an S&S Enrichener, P/N 8519, when installing an E&G/S&S Adapter Kit.

P/N 8512 – ⅝" 93 Big Twin Spacer to convert 92 Big Twin Hypercharger™ to 93 Big Twin. Special Crankcase Breather is also required.

86-87 XL – Not available for stock carb. You must switch to CV carb to use a Hypercharger™.

Electronic fuel injection

Though the carburetors discussed in this chapter are state-of-the-art in fuel efficiency, the new wave of the future will undoubtedly be electronic fuel injection. Electronic injection has been accepted and proven in automotive technology and is considered the only way to go. Even Harley is working on a fuel injection system that will soon debut.

The WhiTek fuel-injection system is unusual in providing a single-fire, computer-controlled ignition that drives two coils, one per cylinder. This ignition system provides any desired ignition curve and is completely adjustable. A handlebar-mounted, liquid crystal display with three integrated control buttons provides all necessary data to perform adjustments. Fuel-air mixture and spark timing *maps* are easily adjusted for maximum performance. Each cylinder is independently adjustable and when changes are noted with the heads, cam, or exhaust, the user is able to make the necessary compensations. The display features rpm readout, battery voltage, and throttle body temperature, and, with optional probes, cylinder and exhaust gas temperatures (Fig. 5-28).

5-28
The WhiTek unit, with electronics under an air cleaner type cap, bolts to the Evo manifold as would a carburetor.

The system delivers increased power, improved gas mileage, and no-lag, instant throttle response, a feature not found with conventional carburetors. The WhiTek system simply bolts on to the standard Bendix-Keihin-style intake manifold. Two sizes are marketed, a 42- and a 46-millimeter. Either will adapt to any Harley Sportster or stroker. The throttle body of the WhiTek unit resembles a velocity stack and it provides uninterrupted airflow to the motor.

A serial port is included in the system to allow control and monitoring from any computer. The electronics are readily accessible and it is simple to swap the existing control program for a new version, allowing software upgrades by the replacement of one memory chip. The WhiTek system is currently available from Rivera Engineering (Figs. 5-29 and 5-30).

5-29
At the rear of the air cleaner throttle body assembly is the manifold mount. This unit bolts to the standard Bendix/Keihin style manifold on almost any Harley motor. A 46-millimeter model is for Big-Twins and a 42-millimeter model for Sportsters.

5-30
The electronics fit between the air cleaner housing and cover.

Accel Thunder-Twin fuel injection

Everyone likes to add power to their Harleys and most of it begins on the fuel-management level. Sophisticated carburetion unfortunately has reached its peak and the future is toward electronic fuel injection. Now it's becoming the way to go for optimum performance from a Harley mill (Figs. 5-31 and 5-32).

5-31
The Accel throttle body injector unit fits easily in the original carburetor area of the Evolution engine.

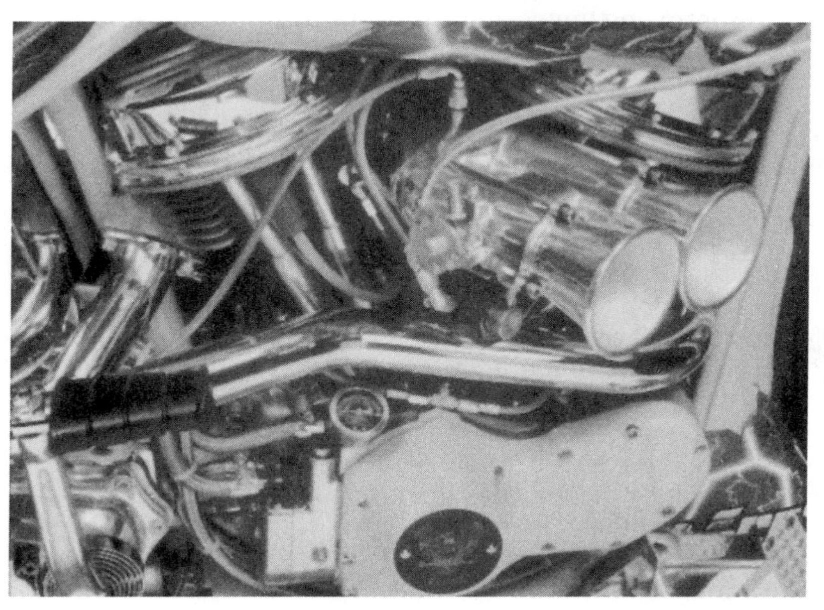

5-32
The Thunder Twin, mounted on Accel's custom demo bike, features a velocity stack style intake. This bike was designed and built by Ben Kudon, custom bike builder and Accel P.R. man.

Accel, backed up by years of research and development, has recently released its revolutionary Thunder Twin fuel-injection system. Accel's new system combines digital fuel injection and electronic ignition control to maximize horsepower and torque, improve throttle response, and provide maximum fuel economy. An efficient engine management system, the Thunder Twin controls the amount of fuel fed to the engine as well as the amount of spark advance. These functions are accomplished digitally in the ECU, or Electronic Control Unit. The ECU constantly monitors operation of the engine via a cylinder head temperature thermocouple and throttle position sensor (TPS) to track throttle movement. A Hall-effect sensor measures engine speed and a manifold absolute pressure sensor (MAP) measures deviations in engine load, or load manifold pressure.

The Accel System is known as a speed-density system. It utilizes engine speed and manifold pressure as primary data sources. The ECU monitors changes in engine speed and manifold pressure and in so doing governs the increase or decrease of fuel going into the engine. The same unit controls timing by monitoring the amount of advance and retard, relative to load and speed experienced by the engine. With the ECU controlling timing and spark, starting the engine is easier than with standard carburetion. Twisting the throttle is not necessary when starting. Turning on the ignition activates the ECU, which informs the injectors of the right amount of fuel to squirt into the intake valves to fire the engine. A continuous fuel flow is then supplied to the engine to maintain an efficient idle.

At the same time, the electronics, by means of the IAC, idle air control, pump air into the throttle body. The IAC acts as the throttle blade in a conventional carburetor, allowing proper passage of air into the venturi to maintain a faster idle to warm up a cold engine. Digitally controlled by the ECU, the idle air control pulls air in by passing it around the throttle plate, allowing the engine to attain a quicker idle without manipulating the throttle. As the engine warms up, the ECU dictates that the amount of air flowing into the throttle body should be decreased by the IAC. When the engine reaches operating temperature the IAC ceases airflow and the engine

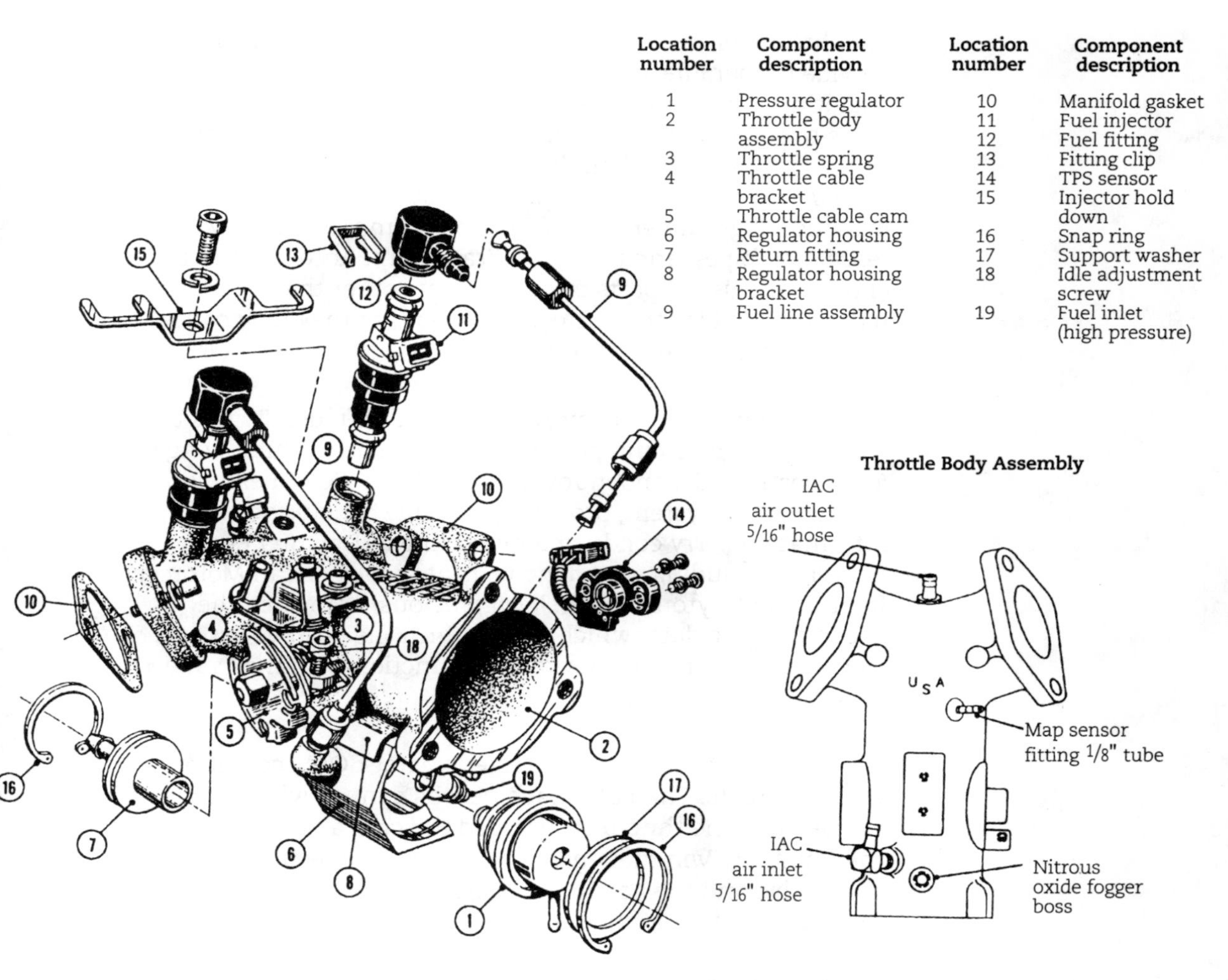

Location number	Component description	Location number	Component description
1	Pressure regulator	10	Manifold gasket
2	Throttle body assembly	11	Fuel injector
		12	Fuel fitting
3	Throttle spring	13	Fitting clip
4	Throttle cable bracket	14	TPS sensor
		15	Injector hold down
5	Throttle cable cam		
6	Regulator housing	16	Snap ring
7	Return fitting	17	Support washer
8	Regulator housing bracket	18	Idle adjustment screw
9	Fuel line assembly	19	Fuel inlet (high pressure)

5-33 *Accel's Thunder-Twin fuel injection throttle body and manifold.* ACCEL Motorcycle Products

idles at a steady rate. Any time the ECU senses a drop in idle rpm it informs the IAC to flow enough air to bring the rpm levels back to the standard 850 to 900 rpm's (Fig. 5-33).

The throttle blade is factory set to coordinate with the IAC to provide proper idle speeds at factory rpm settings at normal operating temperatures. Accel has worked closely with Crane cams to come up with a foolproof, winning, ideal performance street machine. The Thunder Twin system is *dyno* tuned to deliver maximum performance from stock 1984 to 1990 Evolution Big-Twin engines with stock compression, low restriction pipes, and a #1-1100 Crane cam. Accel does not endorse the use of parts or components other than recommended for their street 80 system, as they may cause damage and void the warranty (Fig. 5-34).

Independent testing has proven that the Thunder Twin is the way to go for optimum performance from an Evolution engine. Jerry Dostie, independent owner of a 1991 FXR, did extensive testing on the system, providing valuable conclusive information. Jerry conducted his testing in Daytona Beach, Florida, concluding with dyno tests at the American Motorcycle Institute in Daytona Beach. Jerry graciously provided the accompanying data, which will confirm the positive aspects of the Accel system when used in conjunction with the Crane #1-1100 cam.

First, the machine was dyno tested in stock OEM condition with C.C.I. exhausts. Then the machine was dynoed with the #1-1100 Crane cam. The third test included the cam and the addition of the Thunder Twin fuel injection system. The results are conclusive and the actual dyno sheets are provided for your perusal, substantiating the test results (Fig. 5-35).

The Thunder Twin system also has the capability to run digitally-controlled nitrous oxide. Nitrous oxide injected into the system will give substantial boosts in horsepower and torque. In order to add nitrous you will need an Accel #5557 Nitrous Control kit. Accel systems are available through local Accel automotive dealers, bike shops, or you can obtain them by mail from Rivera Engineering. A definitive manual is provided by Accel, which makes installation of the Thunder Twin system a snap.

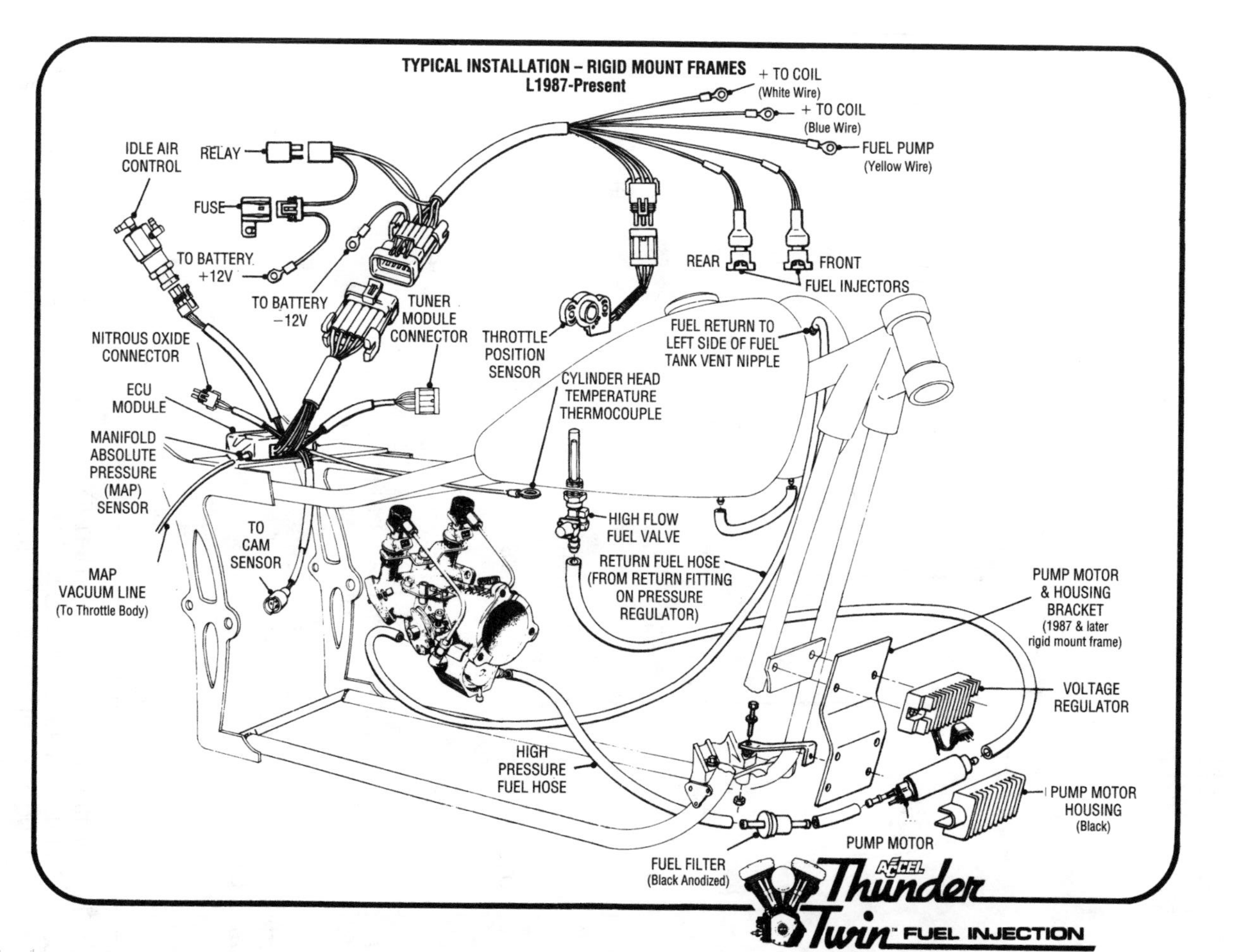

5-34 *Thunder Twin installation details on rigid mount frames from 1987 to present.* ACCEL Motorcycle Products

DOSTIE.013 - H D 1340 FXRP 91 CAM 2-5TH AG 3K
DOSTIE.017 - H D 1340 FXRP 91 CAM 2-5TH AG 3K W/INJ.
DOSTIE.008 - H D 1340 FXRP 91 AS REC 2-5TH AG 2.5K

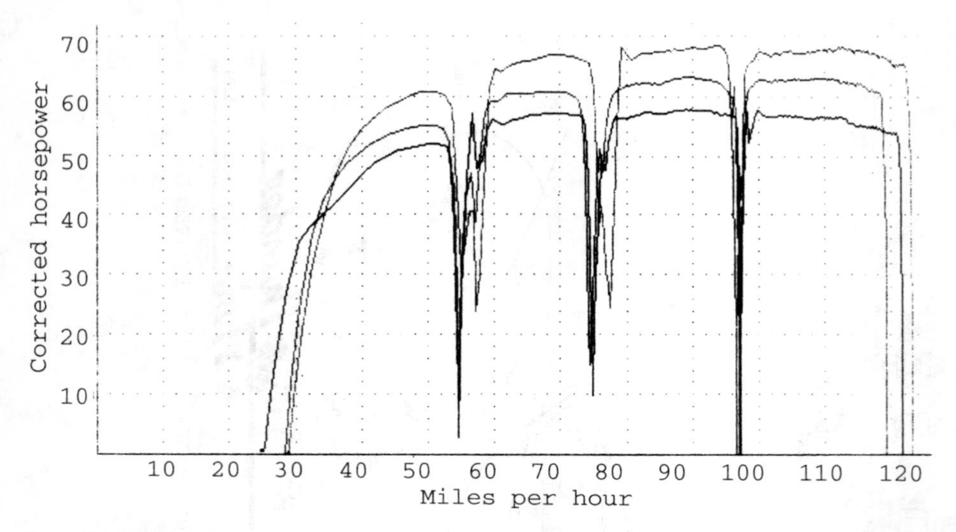

S = 3 As measured on DYNOJET'S MODEL 100 DYNAMOMETER PEP 4.01

DOSTIE.013 79.0 °F 30.33-0.57 in.HG. 0ft., CF = 0.98
1.21.93, JERRY DOSTIE, H.D. FXRP 1340, 1991 MODEL, CCI AFTER MARKET EXHAUST, ALL OTHER STOCK, 7000 MILES ON MACHINE.
CRANE 1-1100 CAM NOW INSTALLED.

DOSTIE.017 67.0 °F 30.13-0.28 in.HG. 0ft., CF = 0.96
2.24.93, JERRY DOSTIE, H.D. FXRP 1340, 1991 MODEL, CCI AFTER MARKET EXHAUST, ALL OTHER STOCK, 7000 MILES ON MACHINE.
CRANE 1-1100 CAM NOW INSTALLED.ACCEL FUEL INJECTOR NOW INSTALLED.

DOSTIE.008 73.9 °F 30.18-0.75 in.HG. 0ft., CF = 0.99
1.13.93, JERRY DOSTIE, H.D. FXRP 1340, 1991 MODEL, CCI AFTER MARKET EXHAUST, ALL OTHER STOCK, 7000 MILES ON MACHINE.SECOND SET OF RUNS

AMERICAN MOTORCYCLE INSTITUTE, DAYTONA BEACH, FLORIDA (904) 255-0295

5-35 *Dyno tests of the Thunder Twin on Jerry Dostie's 1991 FX, conducted at the American Motorcycle Institute at Daytona Beach, Florida. The first test was with all stock components and engine set-up. The second was with the Crane 1-1100 cam, injector, and C.C.I. exhaust installed. The third was on a second set of runs, with the system finely tuned and broken in.*

Nitrous oxide

Turbocharging and supercharging give raw, instant power but are highly expensive alternatives. Harley riders demand instant, foolproof, big power boosts and you can't beat manifold-injected nitrous oxide.

Nitrous is most familiar to us as "laughing gas," mainly used by dentists to reduce the pain connected with dental work. When injected into a carburetor manifold, it's an easy, inexpensive way to pull more horsepower quickly from an engine. What enables the nitrous to provide its awesome power boosts is that it contains 50 to 60 percent more oxygen than normally contained in atmospheric air.

As nitrous oxide is released into the intake system of a Harley engine, transferred by a solenoid-fed electric fuel pump system, it vaporizes and becomes a highly volatile gas. As it changes from a liquid, contained in a storage bottle, it instantly cools to about –130 degrees Fahrenheit, causing the oxygen to become denser. The added density supplements cylinder combustion pressure another 5 to 10 percent. As more oxygen is added, by means of the nitrous, more fuel must also be added to the vaporized mixture to keep it from leaning out. With the common nitrous set-ups, such as the N.O.S. system depicted here, one nozzle feeds nitrous oxide while and adjacent, integrated nozzle feeds in additional fuel. All at the touch of a button, usually somewhere easily accessible on the handlebar.

Low-powered nitrous street systems do not require modification of stock engines. They utilize stock pistons, cams, valves, and ignitions. For all-out competition use, forged pistons are a prerequisite. Nitrous bottles and canisters come in various sizes. The most efficient container is the long, thin N.O.S. unit shown here that will easily mount to frame struts, as in our hardtail frame installation on Felix Lugo's nitrous-fed Sportster. Larger containers will hold up to ten pounds of liquid reserve (Fig. 5-36).

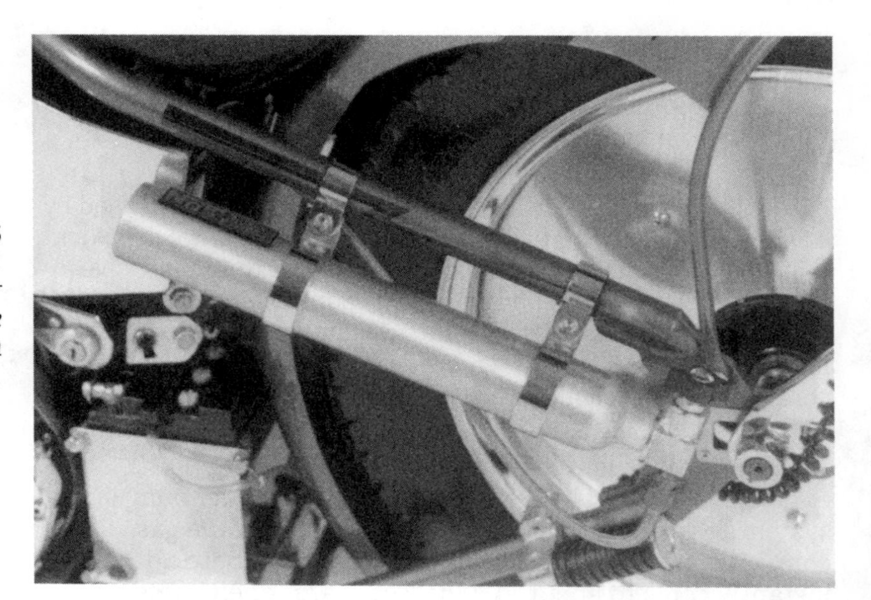

5-36
The long, compact N.O.S. reservoir tank is just right for Harley frame attachment. The unit stores the nitrous in liquid form.

Liquid nitrous feeder lines from the reserve bottle are routed to the carburetor manifold via two solenoid operated valves. Both nitrous and gas are simultaneously fed into the carburetor. On most bikes the horn button is used to trigger the nitrous bursts. The button is modified so that when the system is armed the button provides juice to the solenoid, but not the horn. A separate electrical switch is used to disarm the horn and re-route electrical current to the solenoid. The system is wired so that the solenoid will only be activated when the bike's engine is running (Fig. 5-37).

The engine runs off the carburetor under normal conditions. When you want the acceleration boost you flip the switch, activating the feeder solenoids. When you hit the horn button, ZAP, instant power. Acceleration bursts should be limited to 5 to 10 seconds, enough to leave a non-nitrous competitor in the dust. When pressure in the nitrous tank decreases, built-in safety valves shut the lines to prevent fuel backup. Beware of leaky solenoid valves that allow fuel or nitrous to infiltrate the manifold. Backfiring or errant ignition sparks can cause premature detonation and top end engine damage. Keep in mind that nitrous oxide, when combined with gasoline, is highly volatile.

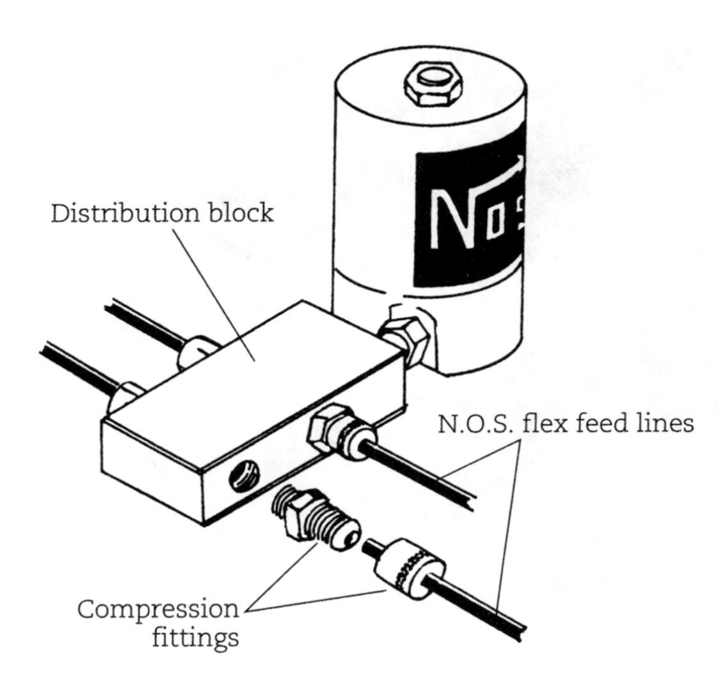

5-37
A solenoid triggers the nitrous circuit and it is fed through a distribution block to the feed lines and to the spray nozzles. A similar circuit, feeding gasoline to the nozzles, is also incorporated in the system.

Another problem that can manifest itself is inadequate fuel volume in relation to nitrous input. Excess nitrous in relation to fuel will cause the vaporized fuel to lean out, promoting heat buildup that can cause damage to piston domes. Another precaution: **Don't hit the nitrous release button until after you have hit the throttle**.

Nitrous installations require little effort but some care, and are easily performed with such basic tools as a wrench, drill, tap, and screwdriver. It's basically a connection procedure except for the manifold, which requires drilling out and tapping of the access orifices for the nitrous/fuel injection nozzles. Holes are drilled with a ¼-inch bit followed by threading with a 1/16-inch taper tap (Figs. 5-38, 5-39, and 5-40).

A dual-feed petcock, from gas tank, with an output nipple to feed gas to the N.O.S. fuel pump is mandatory. The best dual-feed petcocks, providing optimum flow, can be obtained from Pingel Enterprises and work perfectly in nitrous installations. Special brackets for mounting the fuel pump and solenoids are

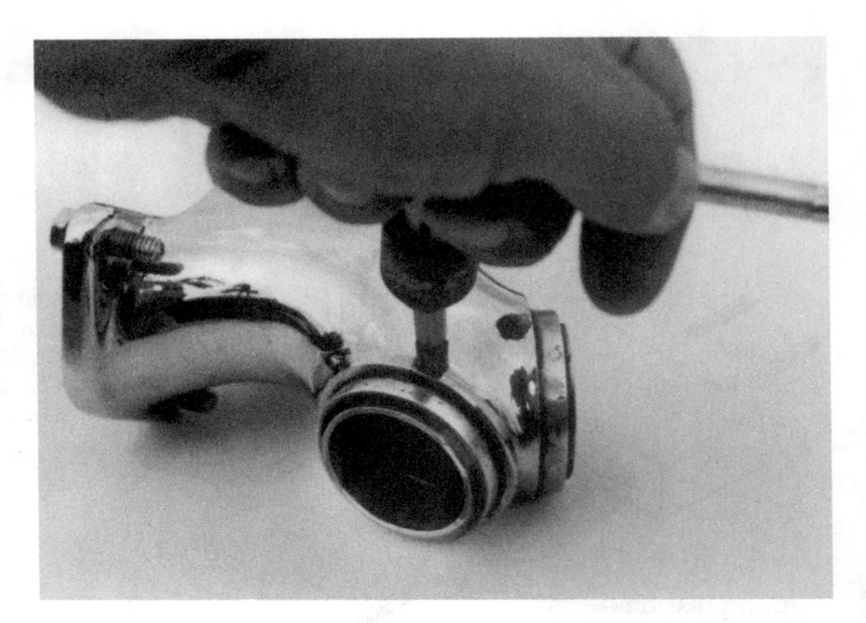

5-38
After drilling, the holes are tapped.

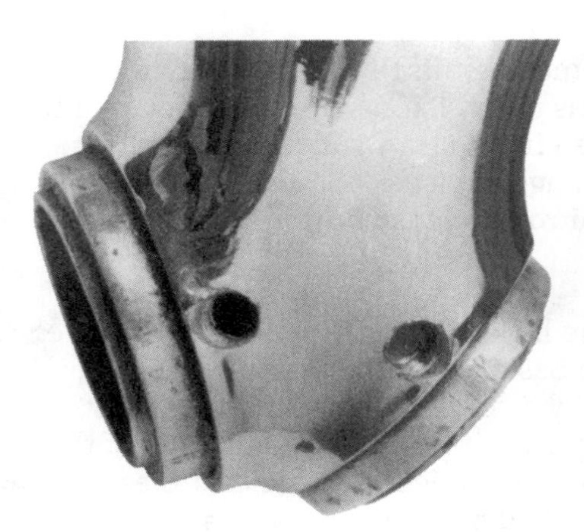

5-39
Tapped holes for the spray nozzles in a Dell'orto manifold.

provided in the Pingel N.O.S. nitrous kits to assist in installation (Fig. 5-41). Solenoids and fuel pump should be mounted apart, but in close proximity to each other.

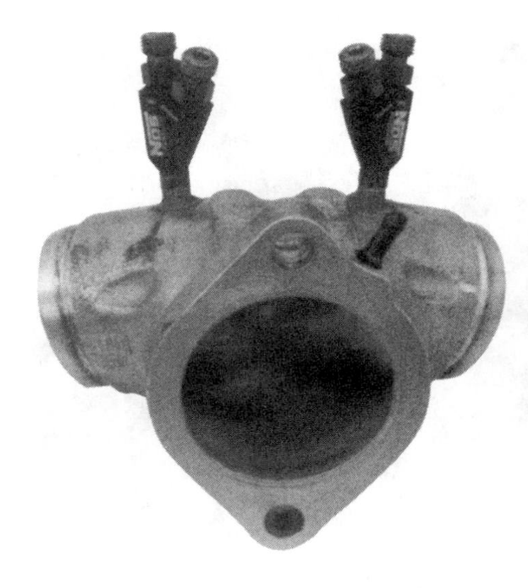

5-40
A pair of N.O.S. nozzles mounted in an S&S manifold. Instructions provided with the Pingel N.O.S. nitrous kit show the proper mounting locations for the injector nozzles.

5-41
The arrow shows the fuel pump feeding gas on Felix Lugo's nitrous Sportster. N.O.S. electrics hide away comfortably under the seat or in various nooks and crannys.

Nitrous spray nozzles mount easily into the tapped orifices on the manifold and are tightened down snugly, but use care not to strip the threads. They are tightened with a 7/16-inch wrench (Figs. 5-42 and 5-43).

5-42
Nozzles can be mounted after the carb and manifold are in place with a 7/16-inch wrench. In virtually all mounting situations, the nozzles will be accessible for servicing.

5-43
N.O.S. nozzles mounted on an S.U. manifold. Note the indication of fuel and nitrous on the body to facilitate line routing.

Turbochargers & superchargers

Both turbochargers and superchargers impart an awesome look to any Harley scooter, but for the average, noncompetitive street rider, they can be highly impractical. Still, some folks swear by the gobs of power they can generate (Figs. 5-44 and 5-45).

5-44
Todd Schuster's Knucklehead benefits from the addition of a Rayjay Turbocharger.

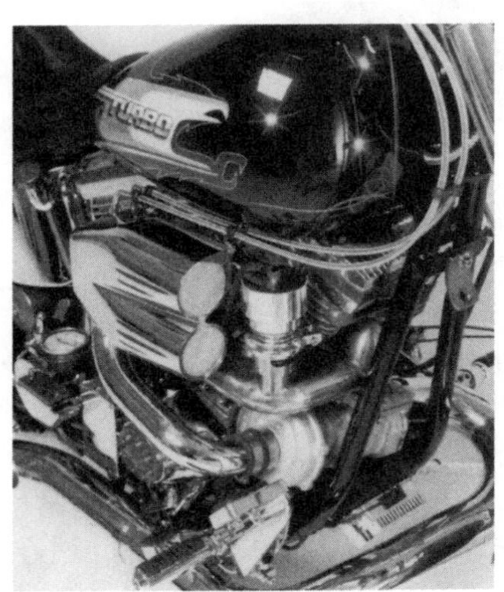

5-45
Turbo Tommy Clark's FXST Evo is not only turbocharged but Nitromethane fuel injected as well.

Turbos

Unless properly set up by an expert, turbochargers are not ideally suited to motorcycles—at least, not street machines. The Japanese adapted them to some production Hondas and Yamahas, but found that they didn't fare very well. The turbo itself is a rotary exhaust-driven compressor, and its primary purpose is to cram more fuel and air into an engine—more than can be sucked in through normal carburetor suction velocity (Fig. 5-46). Car manufacturers usually rely on turbos to generate power boosts, but inherent problems might prevail. When the engine isn't availing itself of turbo feed (air/fuel) into the engine, a small-sized (motorcycle) engine doesn't possess the power necessary to activate the turbo unit. In addition, in order to survive at boosted cylinder pressures, the turbo engine utilizes less compression, somewhere in the neighborhood of 6:1 to 7:1. This tends to make turbo motors sluggish when the booster is low or off.

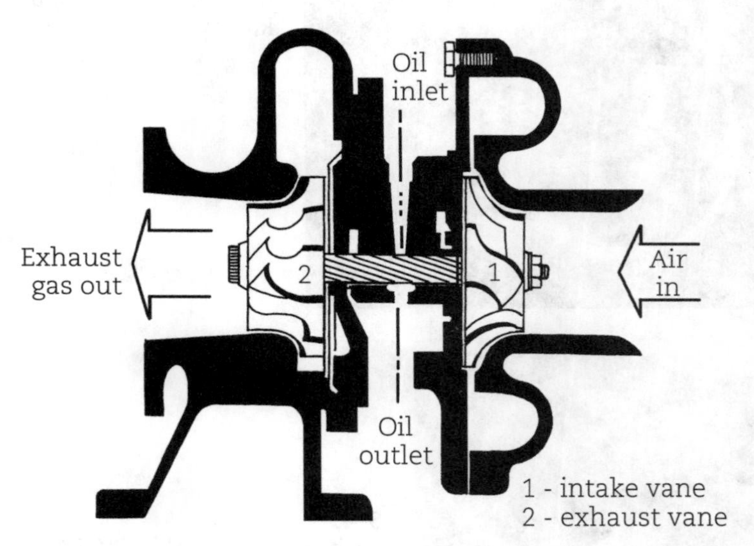

5-46
Cross section of a typical Turbocharger. American Iron Magazine

Motorcycles also happen to magnify the phenomenon known as *turbo lag*, the hesitation period between fuel acceleration feed and the actual manifestation of acceleration punch. Since it is exhaust-driven, the turbo unit takes time to overcome inertia and begin spinning rapidly, thus compressing the fuel/air charge. Many turbos are designed not to spin until mid to high engine speeds, so the lag period is experienced before the turbo accelerates up to an efficiency speed. Turbos are not known for their physical longevity, either. They are driven by hot exhaust gases, which make the unit and its driving impeller bearing very hot (some spin upwards of 35,000 rpm). In addition, Harley engines deliver only two pulses for every two crank revolutions. Since the power pulses aren't spaced evenly, the in-between power gap doesn't assist the turbo in pumping efficiently.

Contrary to the belief that turbos offer free power generated through exhaust flow, there is also back-pressure to contend with (see chapter 6). The back-pressure can impose restrictions on the exhaust system so that the engine must produce a significant added amount of horsepower to force the exhaust through the impeller, leaving the engine with somewhat of a loss of energy that could be used more beneficially to power the motorcycle down the road.

Superchargers

As stated previously, turbos rely on exhaust gases to motivate their turbine impellers; Superchargers differ radically by their more efficient driving systems. Superchargers produce and maintain their power, utilizing engine crankshaft rotation transferred to the blower unit by means of chain, belt, or gear drives. The Supercharger serves the same purpose as a turbo, but delivers more exactingly and efficiently.

The most popular supercharger, and most applicable to motorcycle engines, is the "Roots" blower design (Fig. 5-47). The Roots unit doesn't actually compress the air, but relies on two internally rotating lobed vanes to stuff air and fuel into the engine much faster than the engine could inhale on its own. Superchargers are similar to turbochargers, except that the lobe impellers are driven by the engine's crankshaft.

Roots-type blowers are more commonly seen on bikes. Magnusen blowers, which are out of production and hard to come by, are most coveted as the best supercharging package, partly because they are the easiest to mount on a Harley. Since these Roots-type blowers are crankshaft-driven, displacing air regardless of speed, they provide immediate power boosts devoid of lags inherent with turbos, even at low engine speeds.

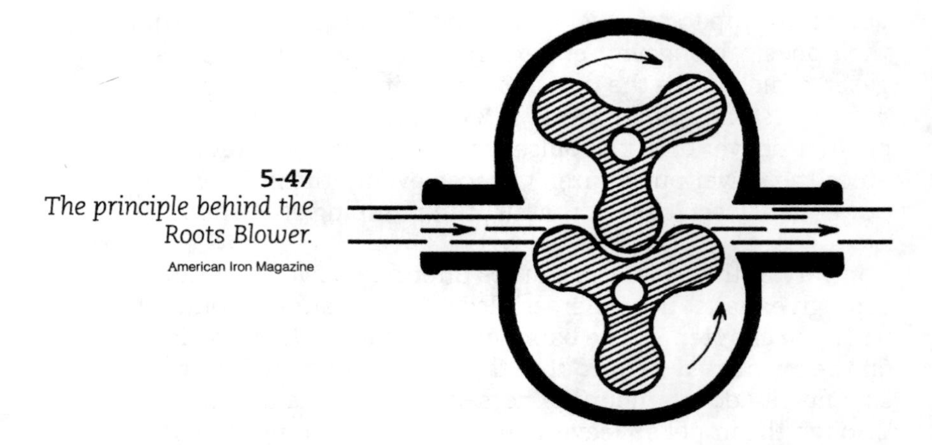

5-47
The principle behind the Roots Blower.
American Iron Magazine

Efficiency might be lost as the Roots spins faster, causing some diminishing of engine power, but its instant boost capabilities and violent acceleration characteristics more than compensate. Though the supercharger may consume up to 10 percent of an engine's power, it serves to increase torque up to about 30 percent.

Turbos and superchargers rely on built-in boost limiting devices in order not to overpower the engine. Turbo units contain a waste gate, which is actually a pressure relief valve opening at a prescribed rev limit. The waste gate allows excess intake to blow either into the exhaust pipe or into the air. Roots units manifest their boost limits by their drive ratios—spelled out as under- or overdrive, related to crankshaft speed. A 10 percent overdrive blower will be rotating at, say, 5,500 rpm, while the crankshaft is spinning at 5,000 rpm. Drive ratio required is governed by how much boost is desired and the displacement of the engine relative to the size and efficiency of the blower.

Backfires are the nemesis of blower-driven engines, especially when intake manifolds are filled to the max with compressed fuel and air. A serious backfire can blow apart the blower case, making for a hairy and costly situation. To circumvent this problem, blowers have built-in pop-off valves such as on heavy-duty compressors or steam engines. The spring-loaded valve "pops" at dangerous pressure levels, relieving pressure which might damage the blower. Even then an excess pressure might build up to the point that will render the pop-off valve inefficient, so added precautions are strongly recommended. Ignition systems should be in A-1 shape, properly timed and right on the money. At high engine speeds, blowers tend to "lean out," which must be compensated for in the carburetor. Leaning-out also manifests itself if the rider backs off the throttle, then goes into jack-rabbit acceleration.

Overall, properly installed and set up, a Roots blower will provide loads of power, if that is what you're looking for and you have an unlimited budget. They are intricate and complex and should be installed only by knowledgeable experienced mechanics. *They are not recommended for average, daily street riding.*

Exhaust systems

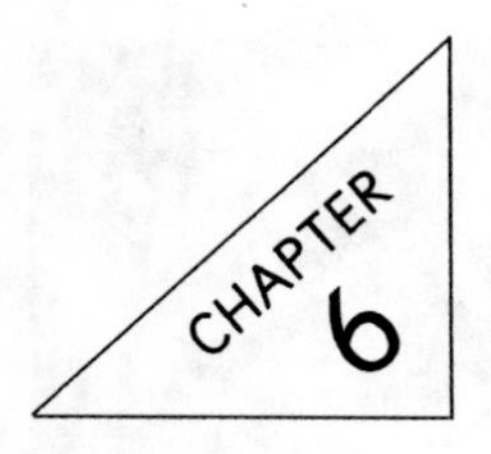

CHOOSING AN EXHAUST SYSTEM requires more than random selection or taking the manufacturer's guidelines as gospel. An exhaust system that performs on a stock bike operates differently when installed on a stroker or beefed-up engine. The object is to coordinate the proper exhaust system with the engine you've tweaked to obtain the performance gains you desire.

Factory exhausts are a compromise intended to result in a smooth-running engine with a minimum of exhaust noise. At the same time, they must function with the cam set-ups and carburation settings. There are also noise and emission requirements that must be adhered to, sometimes at the expense of performance. Some Harley models utilize an unsightly crossover tube. This helps to minimize the restriction of exhaust flow.

Neophytes might jump to the conclusion that 2-inch drag pipes, for instance, offer a solution to the problem. However, in order for unrestricted, wider tubes to work properly, you would probably have to hook them to 95-cubic-inch or bigger re-cammed, high-output motor, which would leave you with high-decibel exhaust noise.

Drag pipes of 1¾-inch diameter are more reasonable, but the stock motor won't generate enough power for them until you're at the valve float level, and again with high noise output. Sticking in baffles tones down the noise but also crimps the power (Fig. 6-1).

Carl Morrow markets True-Flow exhaust pipes that are tuneable. These are the same pipes used by Carl for an unlimited class record at an AHDRA event. The True-Flow exhausts are designed and sold for most all models (Fig. 6-2).

The primary function of the exhaust system is to expel burned fuel and air at the end of the exhaust stroke. This is the point at

6-1
Sumax offers a variety of performance exhaust systems. From top to bottom: flare cut, tapered, slash cut, and drag pipes.

6-2
The same exhaust pipes from Carl Morrow's Git-Kit can be bought separately and used with specific engine set-ups.

which the piston reaches the top of the cylinder chamber, as the intake valve commences opening. The exhaust gas flow is used to start pulling air through the engine, just before the piston begins its intake stroke. If there is too much back pressure or valve overlap, fuel can be sprayed back from the carburetor. Freer flowing exhaust pipes do much to rectify this situation. Conversely, if the pipes are too large, the exhaust flow might not attain sufficient velocity, during valve overlap, to cause sufficient air to move into the carburetor. A proper balance, encompassing all factors, is needed to get the most out of an engine.

Another factor to consider is the *reversion pulse*, which occurs when the exhaust valve is closed and the flow of gases is brought to a standstill. The gases left in the pipe tend to bounce around, up and down the pipe. Correctly utilized, this bouncing reaction can initiate a vacuum at the exhaust valve, timed to the opening of the valve. This assists in getting the flow in motion again. The more back pressure, the less the reversion pulse. There are several factors in obtaining proper reversion tuning: the application of proper pipe diameter, instituting a proper flow rate muffler system, and a pipe length that works best with your particular engine combination.

Some systems incorporate a reversion dam within the exhaust, a step or obstruction in the system that allows smooth exit on the way out but inhibits reverse flow. Reversion dams can be placed in various locations but work best close to the head. The Python units designed by Drag Specialties, for example, have a built-in dam which aids in performance (Figs. 6-3, 6-4, and 6-5).

The nicest thing about an exhaust system, as a performance addition, is that it basically involves only a bolt-on procedure that most anyone can handle.

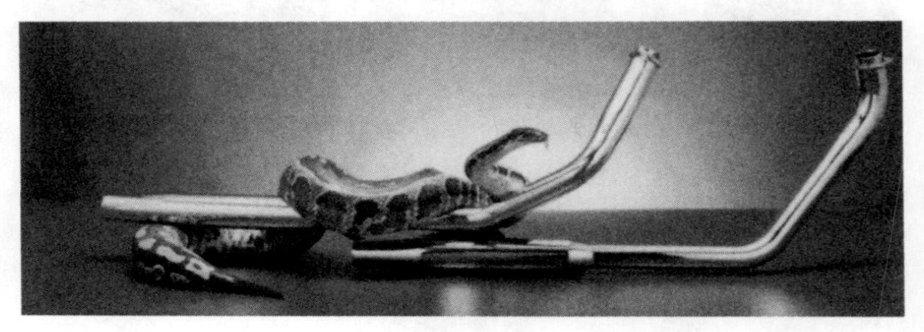

6-3
Drag Specialties Python exhaust pipes.

6-4 *Turnout exhaust pipes by Sumax of heavy-duty, 16-gauge steel, with muffler end tips are for all Evo engines.*

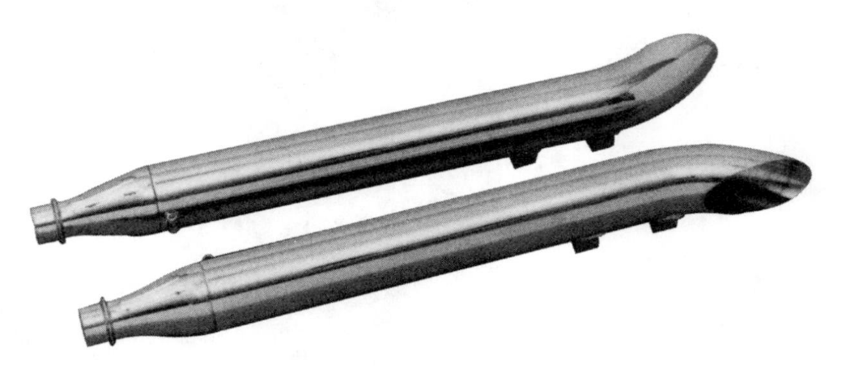

6-5
For those who desire low decibel performance, Drag Specialties feature turnout, or turn-down mufflers.

Remove the pipes and mufflers or end baffles, the frame clamps, and the head fasteners. Install new exhaust gaskets prior to inserting the new pipes into the heads (Figs. 6-6, 6-7, and 6-8). Do not fully tighten the bolts at the heads until the pipes are completely mounted. Then, tighten all the bolts in stages. Finally, slide the mufflers over the pipes and secure them to the frame. Most pipes will bolt-on in stock fashion. In some instances custom clamping, or strapping to the frame, may be required (Fig. 6-9). In this installation, the dual Super Trapp system is used, one of the better aftermarket performance systems.

6-6 *Use exhaust gaskets when mounting a new exhaust system.*

For top performance, it has been proven that a 2-into-1, or *collector*, exhaust is the best way to go. Instead of isolating one pipe from the other, the collector system runs the two pipes into a common chamber. As well as being more compact, collectors offer better muffling. Some feature hollow, baffled cores surrounded by fiberglass or shredded steel strands for increased muffling with minimal performance loss.

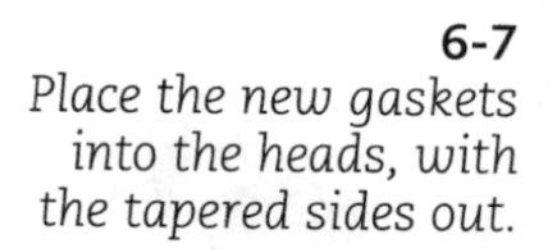

6-7
Place the new gaskets into the heads, with the tapered sides out.

6-8
Insert the pipes and secure them at the heads.

6-9
Install the mufflers onto the pipes and secure them to the frame. All mounting bolts and clamps are then finally tightened.

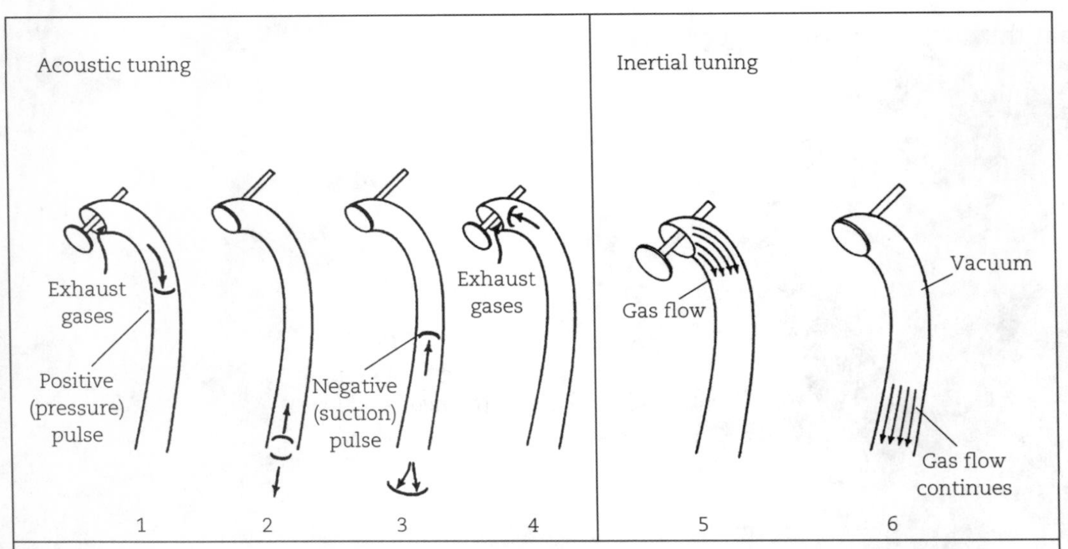

Exhaust Tuning—The Theory Behind Collectors

Most types of tuned exhaust systems achieve horsepower gains by actually providing some suction in the exhaust pipe at the appropriate time (that is, when the exhaust valve opens and gases start to flow out) to suck out more than would exit under normal conditions. You sometimes hear of tuned exhausts that "extract" or "scavenge" the cylinder; these are just fancy words for suck.

There are two ways of achieving this suction—by acoustical tuning and by inerta effects.

The acoustics of exhaust tuning work something like this: the banging of the exhaust valve on its seat as it opens and closes, and the sudden explosion of gases into the pipe produces repeated acoustical waves that travel out the pipe at high speed—and which we hear as beautiful music (custom bike freak) or loud noise (grumpy neighbor) (see Fig. 1). These waves actually travel faster than the exhaust gases that come out. You can see the same effect if you throw a rock into a slow moving river—the ripples will travel much faster than the lazy flow of water downstream. When these exhaust pressure waves finally leave the pipe, they have the curious property of producing a negative (suction) wave that travels back up the pipe (Figs. 2 and 3). If the pipe is the right length, then it is possible to time things for a given RPM so that the negative wave arrives just in time to help suck out another wave as the exhaust valve opens again (Fig. 4). If two or more pipes are joined together this suction wave can be used to advantage in both as long as the lengths of the pipes insure that it returns to each exhaust valve at the right time.

Megaphones intensify this acoustic effect to the point that the engine will give maximum output only over a certain range of engine speed, and runs poorly anywhere else. With straight pipes the difference is not as pronounced, but is spread over a wider range and better for street.

However, what is NOT good for road applications is the amount of noise these acoustically tuned systems produce; and of course any baffling to reduce the racket also seriously impairs the acoustic effects. As a result, unbaffled straight pipes run great. But slip in baffles and your performance may well drop to below stock levels.

This dilemma brings us to the second method of exhaust tuning, which uses the actual flow of the gases and their inertial effects to provide suction. Inertia can be most easily explained by saying that anything—a marble rolling across a table top, or a chopper coasting in neutral up to a stoplight—tends to keep moving unless it is halted by something or someone.

So, when the exhaust valve opens, the heat and pressure in the combustion sends a column of gas down the pipe roaring towards the open end (Fig. 5). At some point along the way, the valve slams closed as the engine goes on to do other things. Since the gases are still moving down the pipe and there is now nothing coming out of the engine to replace them, a partial vacuum is created in their wake (Fig. 6). If the exhaust pipe is long enough so that they are still moving away when the exhaust valve opens again, then a handy suction exists to pull out the next bunch of gases.

Unlike acoustic tuning, this inertial effect can be used for street headers, as long as the muffler used has a big enough straight-through core so that the flow of the gases is not severly impeded. Also, the suction in one pipe can be used to scavenge another if they are connected. For both these reasons, collector exhausts can be inertially tuned to give a happy medium between increased engine output and legal amounts of exhaust noise. The collector-type muffler dampens the acoustical waves (noise), but its large straight-through design leaves inertial effects intact. ●

6-10

One might think that all the pressure forced into the collector might create greater back pressure and power loss, but this is not the case. Harley engines have two cylinders firing per revolution, and the exhaust impulses have enough lag time between them to keep from firing through the muffler at the same time. In other words, each pipe uses the collector at different times, at least in theory (See Fig. 6-10).

Since the system uses only one muffler, the muffler can be physically larger, making it more efficient. The larger size also provides another advantage of the collector, more power with less noise. Horsepower gains of up to 15 percent have been claimed, while keeping exhaust noise well within legal limits.

Most of the top exhaust systems designed today are tuned to extract the maximum from Harley engines. The negative aspect of the collector is the cost, but this is not usually enough to deter the buyer looking for bolt-on horsepower. In addition, collectors lend an appearance of high performance, even to an otherwise stock Harley engine.

Don Rich of Rich Products has designed one of the most efficient, the Rich Thunderheader. It minimizes and redirects the *negative wave* effect of the exhaust gases to increase horsepower throughout the power band (Fig. 6-11).

6-11
The Rich Thunderheader, one of the most effective systems around, designed for top performance.

The Rich system incorporates two main components: an internal flow director and an outer reverse cone welded over an inner reverse cone. The cone works as a diffuser to isolate the reverse wave and vent it into the air, which creates a low pressure area in the exhaust that assists in pulling gas into the combustion chamber. It's similar to an expansion chamber on a 2-stroke engine. Tested on a flow bench, the Rich system has been found to dispel 22 percent of the negative wave, with commensurate horsepower gains. The system works ideally on street machines, but it is loud.

The popular Super Trapp exhaust systems for Harleys are also excellent high-performance collectors, particularly for FXRS Evo motors. Offered in stainless steel or chrome, Super Trapp's system allows for precise tuning by adding or subtracting diffuser discs inside the rear baffle.

The Super Trapp 2-into-1 system is not only tops in looks, but is an optimum performer. One of the best features of the Super Trapp is its tuneability; noise and performance are adjustable, and the tuning can be accomplished with one allen wrench. The system is supplied with diffuser discs, up to 12 depending on the model. Using four discs will flow the same amount of exhaust as a stock system. Using up to eight discs will allow operation with stock jetting. Using more than eight discs will require jetting changes. By adding discs three things are accomplished: the exhaust noise increases, the motor runs leaner, and the horsepower and power band is increased. The opposite happens as the discs are removed.

Jetting

You may mount a new collector but find that, after installation, it still doesn't allow your engine to perform up to snuff. The motor depends upon a proper gas-to-air ratio to perform optimally. Changing the airflow requires that you adjust the gas flow as well. If a proper balance is not realized, you will experience engine bogging and skipping, carbon buildup, and, worst of all, a loss in horsepower. To compensate, you must make changes in carburetor *jetting*.

You can raise or adjust the needle in the jet. Raising the needle allows greater gas flow, lowering the needle decreases it.

Needles are available in a number of design tapers too, which serve the same purpose. You can also change to a larger jet, which provides more gas and richer mixtures.

The next jet most likely to be changed is the pilot jet, usually located off to the side of the needle jet. Changes in the pilot jet will affect low end performance. Also to be considered is the air correction jet, which only acts as a high speed jet. This is only to be tinkered with if the engine tends to bog at full throttle. If you are instituting radical exhaust changes in your bike, it is feasible to jump one to three jet sizes. Under normal circumstances, when trying to correct a lean or rich condition, try to pinpoint in what range the problem manifests itself and select a needle or jet for that range. Then start moving up or down, a step at a time, until the appropriate setting is reached.

CHAPTER 7

Brake systems

STOCK HARLEY BRAKE SYSTEMS offer adequate stopping power and cannot be faulted for efficiency. Disc brakes, the norm on all contemporary motorcycles, provide the most stopping power. The drum brake is a thing of the past. The Harley OEM units are mass produced and not as sophisticated as many aftermarket offerings. Performance oriented builders of highly powered engines may require more stable, faster acting, and more controllable brake systems. These are available from aftermarket producers who offer precision machined, billet-aluminum, state-of-the-art products. They also offer dual disc capabilities, the ultimate in stopping power (Fig. 7-1).

7-1
For quick acting stopping power dual discs are the ultimate.

Fortunately aftermarket disc brake packages are bolt on items. Standardization in mounting is the norm and you can fit virtually any aftermarket brake caliper onto any front or rear end, as you would a stock unit (Fig. 7-2).

7-2
Performance Machine offers brake systems for any stock type Harley mounting.

Performance Machine brake systems are sensitive and controllable, yet not grabby. The more pressure necessary to activate the brakes, the more difficult it is for the rider to ascertain what the system is doing. The ideal degree of sensitivity would be a two-finger front brake, requiring the least amount of hand movement. Disc brake components need to be light but strong enough not to flex under extreme braking conditions. Billet aluminum is the material of choice for high performance brake components and it works well under high stress conditions. Performance Machine offers an array of calipers in two-, four-, and six-piston configurations and accompanying brake rotors, in conventional sizes (Fig. 7-3).

The Performance Machine six-piston, differential-bore caliper, the 137×6, was designed for extreme braking efficiency. The six piston, dual-pad design increases pad contact on the disc and provides excellent piston retraction, for drag-free operation. The differential piston bore sizes compensate for uneven pad wear, at high operating temperatures, by exerting differential forces at the leading and trailing edges of the pad, resulting in more uniform pad wear. The pads are also thicker than conventional, four-piston types. The 137×6 is excellent for street Harleys with single disc systems that could stand improved

7-3
The Performance Machine 125x4R four-piston caliper in a Harley swing arm set-up.

7-4
The Six-Piston 137x6 Performance Machine differential bore caliper and one piece disc will bolt onto most Harleys with appropriate brackets.

stopping power. The six-piston caliper allows dual disc performance and efficiency with single-system weight (Fig. 7-4).

F.C.C. produces a caliper that is a more moderately priced alternative to billet aluminum. These are cast aluminum units, designed to meet rigid racing standards, with the ability to accept any diameter and thickness rotor. All DOT-rated brake fluids are compatible with these calipers and they meet or exceed most OEM specifications.

All F.C.C. calipers and Harley mounting brackets are highly polished or chrome-plated. Complete kits are available for either top or bottom mounting. Dual sets of calipers can be mounted on a single rear disc, for eight-piston caliper stopping power (Fig. 7-5).

7-5
The F.C.C. caliper is a cast aluminum alternative to machined billet and economical as well as efficient.

Storz Performance imports the Grimeca brake calipers, made in Italy. The Grimecas are small and lightweight and feature Ferodo pads. The tiny 1000T Grimeca caliper weighs only 15 ounces and uses a 30-millimeter piston, making it one of the smallest calipers available. It is non-sided and can be used in either right or left hand applications. Slightly larger, the 1025 model features 4-millimeter pistons in a tight body configuration. For the stopping power of dual piston calipers, Grimeca markets their 1050/1055 unit for either right or left hand applications. The dual piston units have mounting holes on 3½-inch centers and weigh only 2 pounds, 4 ounces.

Sumax has available the Jay-Brake line of quality brake components which will adapt to all front and rear Harley disc brake set-ups (Fig. 7-6). The Jay-Brake calipers are of high-quality billet construction and feature 1.25-inch-diameter steel pistons for durable and efficient operation. Three models are provided: the DC-88 dual piston unit, the QC-88E double dual piston unit with ears, and the QC-88 double dual type without ears (Fig. 7-7).

7-6
The Jay-Brake dual piston unit with ears, mounted on a softail Harley.

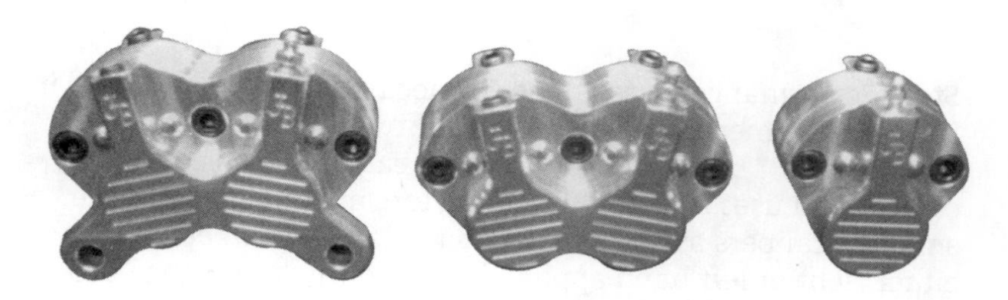

7-7
Jay-Brake Calipers left to right: The QC-88 Dual (4) piston unit with ears; the QC-88 Double Dual without ears; and the Dual DC-88 two piston unit.

For rear disc brake operation forward controls are offered by Jay-Brake and Performance Machine. Forward controls are designed with a pivot point close to the footpeg so that you can brake without removing your foot from the forward mounted peg. Matching forward shift controls are also marketed (Figs. 7-8 and 7-9).

7-8
The Performance Machine forward brake control locates the brake cylinder, front peg, and the brake lever all in close proximity.

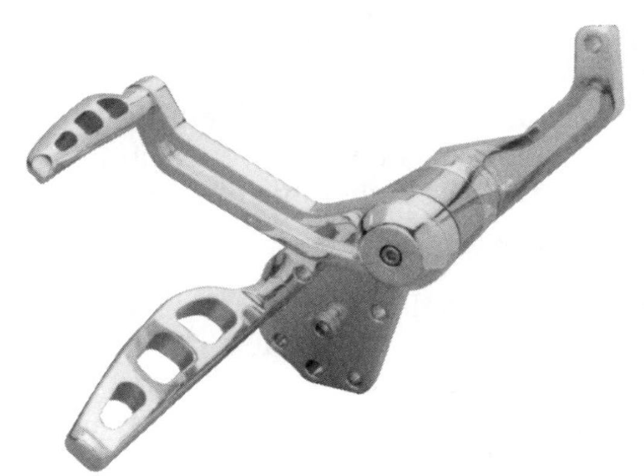

7-9
The matching P.M. forward shift unit.

Brake pads

Only the highest quality brake pads should be used to obtain optimum, fade-resistant braking. The new Kevlar disc pads, recently made available by EBC Brakes of England, use the latest friction material technology to provide the best stopping power with the least noise, fade, or rotor wear. Kevlar, a man-made fiber produced by Dupont, is six times as strong as steel, yet resilient enough to be an ideal brake pad material. Kevlar carbonizes at over 1000 degrees, much higher than previous fibers used by brake pad manufacturers. These are far superior to conventional asbestos pads.

Brake fluids

There is a difference between DOT 4 brake fluid, used in older brake systems, and the DOT 5 used in most modern motorcycles. To understand the differences, you must visualize what happens when you press your brake pedal, or lever, and activate the hydraulics. As you depress the brake lever, you transfer fluid into the caliper pistons, which in turn squeeze the pads against the rotor, causing the friction which slows the bike down. The more the brakes are applied, the hotter the fluid gets, which produces moisture.

If there is no moisture in the system, DOT 5 does not reach boiling point until 700 degrees Fahrenheit, whereas DOT 4 will begin to boil at 446 degrees. That is a vast difference thermally. The two fluids also differ physically and in appearance. DOT 5 is purple in color and DOT 4 is sand colored. DOT 5 is more slippery than DOT 4. DOT 5 is a silicon-based synthetic fluid while DOT 4 is a Polyglycol, an organic fluid. DOT 5 is a better lubricant, affording the internal parts longer life. The DOT 5 will maintain an almost constant viscosity throughout the operating temperature range of the system. DOT 4 acts in the same manner as motor oil. It thins out when hot and thickens when cold.

In a DOT 5 system, braking action is the same in sub-zero weather as in high heat conditions. DOT 5 will absorb only .028-percent water, by weight, from the air. DOT 4 can absorb up to 6 percent and have its boiling point lowered to 311 degrees. As moisture enters the system, the boiling converts the water into a gaseous state that creates air in the system. As you continue to brake, the fluid rises in temperature. The hotter the fluid gets, the more air manifests itself and the less brakes you have. This is why the more you use your brakes, the less they seem to react. As the air compresses, you get that spongy feeling known as brake fade. Unfortunately, brake fluid tends to readily draw in moisture from the air. Water can easily be drawn into the brake system from the air reservoir in the master cylinder or through the brake lines.

Rumors have surfaced to the effect that DOT 5 will damage gaskets and seals. If you convert from DOT 4 to DOT 5 you probably will get some leaks, but not due to the DOT 5. Since DOT 5 is thinner and highly lubricating and if seals aren't in good shape or pistons are worn, the DOT 5 will seep through. Synthetic fluid has better lubricating properties than natural oil, but it will escape if seals are not in top shape. To go the DOT 5 route conditions must be ideal before you make the swap. The best time is when you overhaul the brake system; bleed out the old fluid and replace the seals. If you upgrade from DOT 4 to DOT 5, plan on rebuilding the master cylinders and replacing the rubber brake lines. DOT 5 will provide you with less maintenance and improved braking performance. When you switch, discard the old brake hoses in favor of new braided-steel lines, which don't flex and wear twice as well.

Harley gallery

The master himself, Arlen Ness, entrepreneur of the "Superbike," with son Cory, judging at the Fourth Annual Brothers III Anniversary Bash.

Ace Armstrong with his custom cafe racer. This Harley 1991 Sturgis features Fueling-Rivera Four-Valve heads and was built by Phil Petersons Harley-Davidson, Miami, Florida.

Doug Doan's 1991 FXRS built by owner and painted by Joe Spadaro. This bike features Rich headers and Andrews EV-3 cams for added power.

John Sachs, engine specialist, built this Sportster street beater around a 1986, stroked, 89-cubic inch motor. Construction and refined enginework by Sachs, himself.

Felix Lugo's "Intimidation" 1971 Sportster relies on a NOS nitrous oxide system for power acceleration bursts.

Methane-injected turbo engine system on Turbo Tommy's street killer was designed, modified, and built by master engine builder, Scott Baringer.

Designed and built by Custom Accessories, Pompano Beach, Florida, Turbo Tommy's 1984 FXST has a modified 1990 Evo engine.

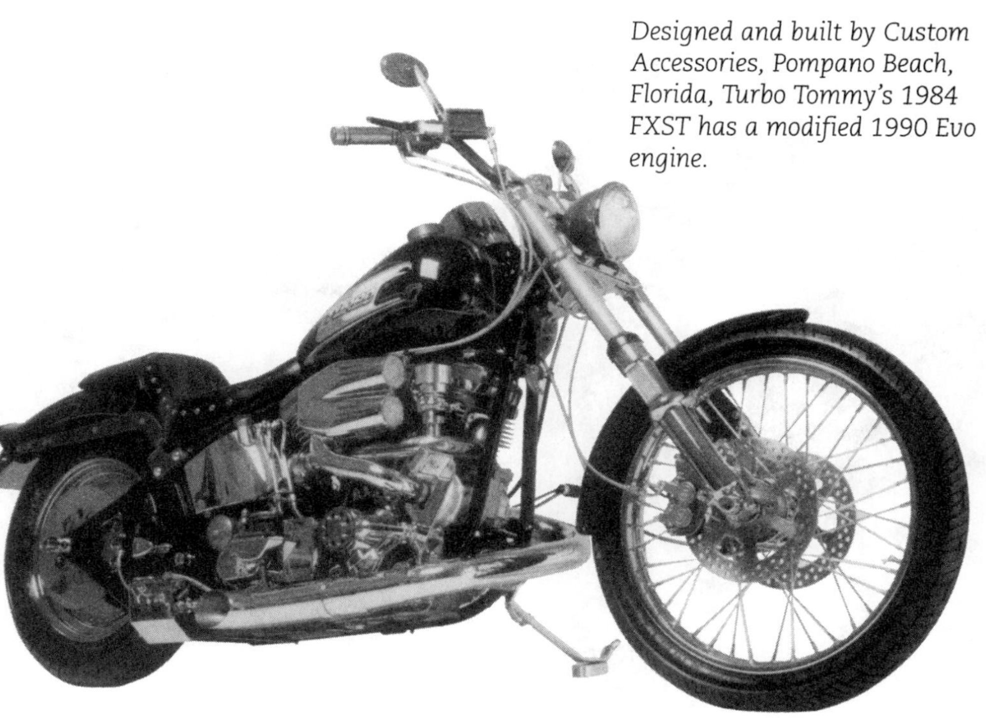

Dan Giaraglione's 1986 FXST show-and-go classic.

"Terminator" 1987 FXSTC was built by owner Dave Amchir and Custom Accessories, Pompano Beach, Florida. Enginework by Scott Baringer.

Mark Shadley's Supercharged Sportster was built by Dave Perewitz, Brockton, Massachusetts.

Lou Falcigno's "Overbored" high-output Shovelhead early 80s street chopper is done over in 60s styling.

Jack Rouse's late FXST features dual Dell'Orto carbs (one on each side fed into the specially designed manifold) and Indian-styled fenders for a novel look. Bike built by Custom Accessories, enginework and modifications by Scott Baringer.

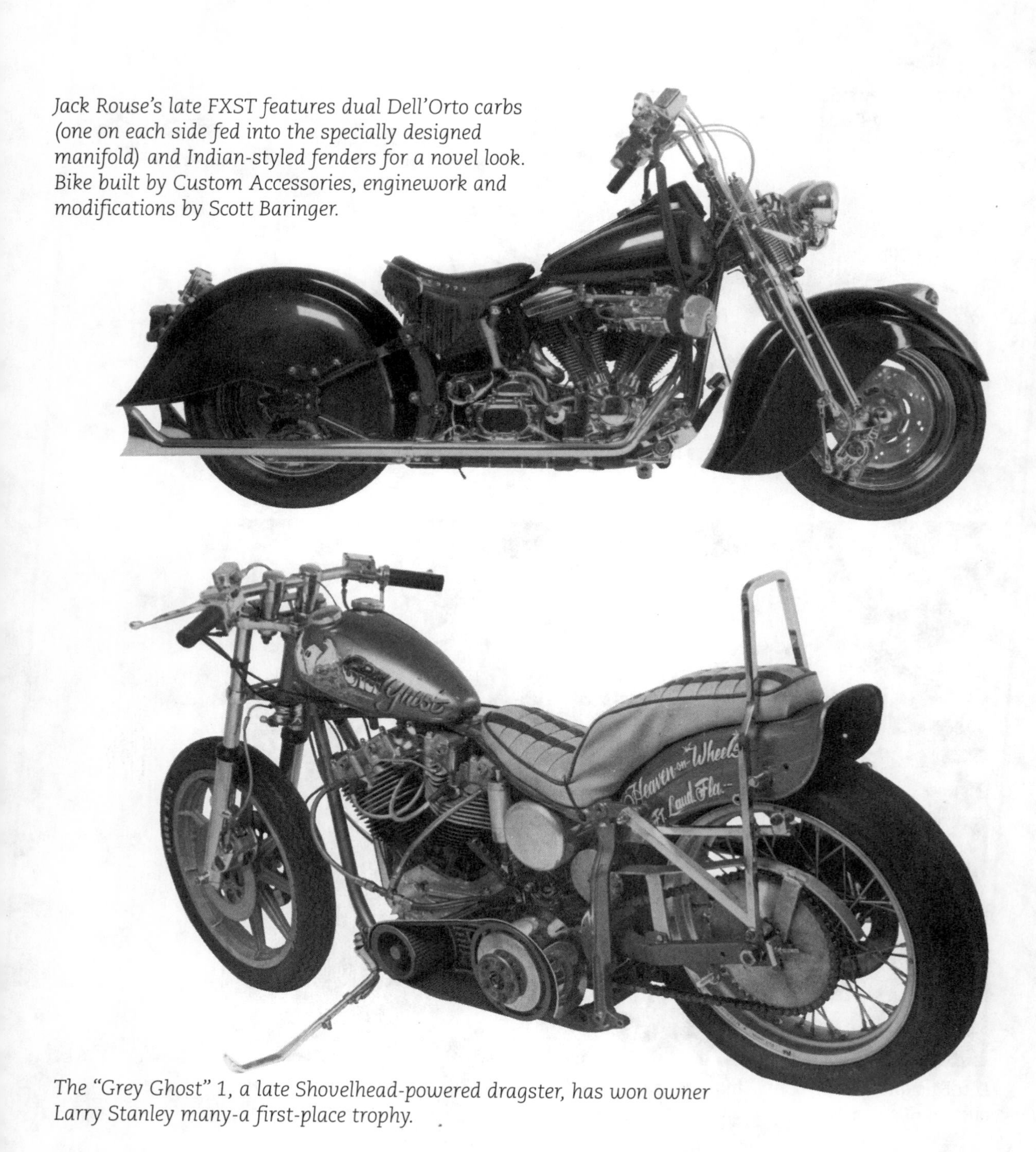

The "Grey Ghost" 1, a late Shovelhead-powered dragster, has won owner Larry Stanley many-a first-place trophy.

Custom Dual Dell'Orto custom-painted and engraved carb.

Dave Amchir's 1987 FXSTC built by Dave and Custom Accessories, Pompano Beach, Florida. (For another view, see page 192.)

Jack Rouse's 1990 FXSTS built by Scott Baringer of Custom Accessories, Pompano Beach, Florida. This bike features two dual Dell'Ortos, one on each side. (See color section for side view.)

Carl Morrow's Bonneville Streamliner built around a late-model XL engine. Morrow's Sportsters hold more world records for speed than most contemporary contenders.

Outstanding example of refined motorwork as seen at a custom show.

APPENDIX B

Suppliers

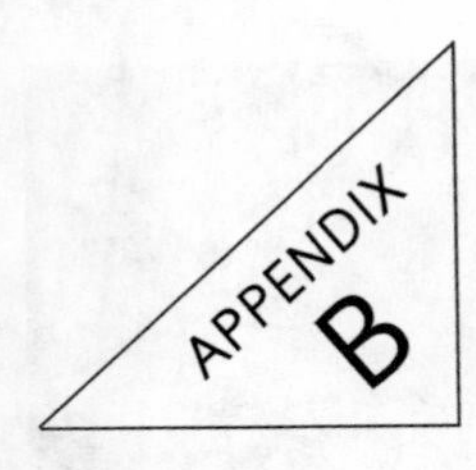

Carburetors

Custom Chrome
1 Jacqueline Ct.
Morgan Hill, CA 05037

Drag Specialties
9839 W. 69th St.
Eden Prairie, MN 55344

Rivera Engineering
6416 S. Western Ave.
Whittier, CA 90606

S&S
Box 215
Viola, WI 54664

Cams

Andrews Products, Inc.
5212 Shapland Ave.
Rosemont, IL 60018

Carl's Speed Shop
9339 Santa Fe Springs Rd.
Santa Fe Springs, CA 90670

Crane Cams
530 Fentress Blvd.
Daytona Beach, FL 32014

Rivera Engineering
6416 S. Western Ave.
Whittier, CA 90606

Sifton MC Products
943 Branston Rd.
San Carlos, CA 94070

Disc brakes

Jay Brake Ent.
211 Grant St.
Lockport, NY 14094

Performance Machine
15220 Illinois Ave.
Paramount, CA 90723

Frames

Arlen Ness Ent.
15997 E. 14th St.
San Leandro, CA 94578

Paughco, Inc.
11 Cowee Drive
Carson City, NV 89701

Sumax, Inc.
337 Clear Rd.
Oriskany, NY 13424

Front ends

Paughco, Inc.
11 Cowee Drive
Carson City, NV 89701

White Bros.
14241 Commerce Drive
Garden Grove, CA 92643

Storz Performance
1445 Donlon St.
Ventura, CA 93003

Hi-performance—big bore cylinders

Axtell
1424 S.E. Maury
Des Moines, IA 50317

Hyperformance
512-A N.E. 12th Ave.
Pleasant Hill, IA 50317

Rev-Tech
1 Jacqueline Ct.
Morgan Hill, CA 95037

S&S
Box 215
Viola, WI 54664

Sputhe Engineering
11185 Lime Kiln Rd.
Grass Valley, CA 95949-9715

Trock Cycle Specialties
13 N 417 French Rd.
Hampshire, IL 60140

Ignition systems

Accel
175 N. Branford Rd.
Branford, CT 95742

Crane Cams
530 Fentress Blvd.
Daytona Beach, FL 32014

Dyna-Tech
810 N. Cummings Rd.
Covina, CA 91724

Morris Magneto's
103 Washington St.
Morristown, NJ 07960

Rivera Engineering
6416 S. Western Ave.
Whittier, CA 90606

Screaming Eagle
3700 W. Juneare Ave.
Milwaukee, WI 55201

Pistons

Axtell
1424 S.E. Maury
Des Moines, IA 50317

Rivera Engineering
6416 S. Western Ave.
Whittier, CA 90606

Replacement engine cases (Big-Twin)

Delkron
2430 Manning St.
Sacramento, CA 95815

S&S
Rt. 2, County G
Viola, WI 54664

Sputhe Engineering
11185 Lime Kiln Rd.
Grass Valley, CA 95949-9715

S.T.D. Development
P.O. Box 3583
Chatsworth, CA 91313

Stroker (bottom end) components

S&S
Box 215
Viola, WI 54664

Transmission cases

Sputhe Engineering
11185 Lime Kiln Rd.
Grass Valley, CA 95949

S.T.D. Development
P.O. Box 3583
Chatsworth, CA 91313

Transmission gears

Andrews Products, Inc.
5212 Shapland Ave.
Rosemont, IL 60018

Top-notch engine work

Valve work, porting and polishing Big Bore, stroked engine, etc.

Carl's Speed Shop
9339 Santa Fe Springs Rd.
Santa Fe Springs, CA 90670

A

B

C

About the Author

Carl Caiati is an professional writer and photographer as well as a custom automobile painter and airbrush specialist. His articles appear frequently in *Hot Rod*, *American Iron*, *Popular Mechanics*, and *Motor Magazine*. He is the author of TAB/McGraw-Hill's *Customizing Your Harley*.